PIVOTING AND EXTENSIONS:
In honor of A.W. Tucker

MATHEMATICAL PROGRAMMING STUDIES

NORTH-HOLLAND PUBLISHING COMPANY – AMSTERDAM · OXFORD
AMERICAN ELSEVIER PUBLISHING COMPANY, INC. – NEW YORK

Pivoting and Extensions:

In honor of A.W. Tucker

Edited by M.L. BALINSKI

I. Adler
M.L. Balinski
R.W. Cottle
G.B. Dantzig
R.J. Duffin
B.C. Eaves
D.R. Fulkerson

J.H. Griesmer
A.J. Hoffman
H.W. Kuhn
S. Maurer
K.G. Murty
R. Oppenheim
L.S. Shapley
P. Wolfe

1974

NORTH-HOLLAND PUBLISHING COMPANY – AMSTERDAM · OXFORD
AMERICAN ELSEVIER PUBLISHING COMPANY, INC. – NEW YORK

This book is also available in journal format on subscription.

Library of Congress Catalog Card Number : 74-83270

North-Holland ISBN for this Series : S 0 7204 8300 X
for this Volume : 0 7204 8301 8

American Elsevier ISBN : 0 444 10727 4

Published by :

North-Holland Publishing Company – Amsterdam
North-Holland Publishing Company, Ltd. – London

Sole distributors for the U.S.A. and Canada :

American Elsevier Publishing Company, Inc.
52 Vanderbilt Avenue
New York, N.Y. 10017

PREFACE

This volume initiates the series MATHEMATICAL PROGRAM-
MING STUDIES. Conceived as a companion to the journal MATHE-
MATICAL PROGRAMMING, each STUDY is to be centered about
a unifying subject matter and to consist of either a collection of papers,
a single monograph, the proceedings of a specialized symposium, a
guide to computational practice, or any clearly focused and useful
piece of work.

The conjunction of the launching of the STUDY series and the retirement from Princeton University of A.W. Tucker proved to be too
neighborly to withstand the logic, sentiment and pleasure of combining
the events to honor him. Accordingly, his friends, colleagues, students
and sons were invited to add their contributions to a volume devoted
to an area pioneered and vigorously promoted by A.W. Tucker throughout the last decade: PIVOTING AND EXTENSIONS. His pervasive influence on the field of mathematical programming, and on the
people who work in it, is clearly displayed by the contents of this volume.
The affection and respect for him is explicit in this volume, but was
also expressed in the many letters of regret received from invited con-
tributors who were precluded by the subject matter from submitting
their own papers. There is still another fitting coincidence: this STUDY
series carries further an idea initiated by Professor Tucker which led
to the establishment of the Annals of Mathematics Studies.

Pivoting is an essential computational and theoretical tool in mathematical programming. Although based on the Gauss–Jordan complete elimination or replacement idea of linear algebra, its computational significance was only fully realized with the development of the
simplex method for linear programming in 1947. Since then, the use
of the pivoting idea in theory and computation has proliferated and has
led to developments in linear, nonlinear and integer programming, the
structure of convex polytopes, matrix and bi-matrix games, the computation of fixed points and economic equilibria, the complementarity
problems, and still other domains. The papers of this STUDY, many of
which were presented at the 8th International Symposium on Mathe-

matical Programming held at Stanford University in August 1973, bear witness to this statement.

The first four papers deal with the structure of convex polytopes. Paper 1 establishes improved lower bounds for the maximum diameter of polytopes. Papers 2 and 3 introduce and study "abstract polytopes", a combinatorial construct which generalizes the (pivotal) structure of extreme points and their adjacencies but is a particular type of pseudo-manifold. In Paper 4 the Hirsch conjecture is established constructively for two classes of transportation polytopes.

Several papers are related to the complementary pivoting ideas first introduced by Lemke and since extended in many directions. Paper 5 characterizes the existence of a solution ray to a linear complementarity problem when the "structured" matrix is copositive plus. Paper 7 describes an algorithm for solving piecewise linear convex equations. Paper 10 provides an algorithm which proves the fundamental theorem of algebra by building a labeling procedure, pivotal in character, upon a combinatorial lemma concerning labeled triangulations of the complex plane. Paper 12 gives an expository account of the original Lemke–Howson idea in terms of a geometric labeling system for bi-matrix games and presents an orientation theory for the equilibrium points and the complementary pivot paths connecting them for such games.

Paper 6 modifies Fourier's original method of elimination of variables for solving systems of linear inequalities thus obtaining a practical computational algorithm for a class of parametric linear programs.

Paper 8 shows that optimal integer solutions exist to certain types of linear programs given a "balanced" structural matrix and nonnegative integer right-hand sides, thus generalizing previous results of Berge, by using the theory of blocking and anti-blocking pairs of matrices.

Paper 9, an application to coding theory, uses integer programming cuts and pivoting to establish a lower bound on the uniform length of word required for a binary linear error-correcting code satisfying a minimum distinguishability criterion.

Paper 11 uses Tucker's pivotal algebra to introduce determinants in a new, simpler and computationally efficient way.

Finally, Paper 12 presents a complementary pivot algorithm for the problem of finding the point of a polytope in Euclidean space having smallest Euclidean norm, a special form of quadratic program which admits a special algorithm having a transparent geometry.

M.L. Balinski

CONTENTS

Preface v
Contents vii

Albert William Tucker 1
Bibliography 4
In honor of A.W. Tucker's contributions to mathematical
programming, *George B. Dantzig* 9

(1) Lower bounds for maximum diameters of polytopes, *I. Adler* . 11

(2) Maximum diameter of abstract polytopes, *I. Adler and G.B.
Dantzig* 20

(3) Existence of *A*-avoiding paths in abstract polytopes, *I. Adler,
G.B. Dantzig and K.G. Murty* 41

(4) On two special classes of transportation polytopes, *M.L. Balinski* 43

(5) Solution rays for a class of complementarity problems, *R.W.
Cottle* 59

(6) Fourier's analysis of linear inequality systems, *R.J. Duffin* . . . 71

(7) Solving piecewise linear convex equations, *B.C. Eaves* 96

(8) On balanced matrices, *D.R. Fulkerson, A.J. Hoffman and R.
Oppenheim* 120

(9) Derivation of a bound for error-correcting codes using pivoting
techniques, *J.H. Griesmer* 133

(10) A new proof of the fundamental theorem of algebra, *H.W. Kuhn* 148

(11) Pivotal theory of determinants, *S. Maurer* 159

(12) A note of the Lemke–Howson method, *L.S. Shapley* 175

(13) Algorithm for a least-distance programming problem, *P. Wolfe* 190

Albert W. Tucker

Mathematical Programming Study 1 (1974) 1–3. North-Holland Publishing Company

ALBERT WILLIAM TUCKER

Albert William Tucker was born in Oshawa, Ontario, Canada on 28 November 1905. After earning B.A. and M.A. degrees at the University of Toronto, he went to Princeton University where he finished his Ph.D. in mathematics in 1932. Thus began an association with Princeton that has continued throughout his academic career. He joined the Princeton Faculty in 1933, was progressively promoted, and in 1954 succeeded Emil Artin as Albert Baldwin Dod Professor, which post he held until his retirement in June 1974. He was Chairman of the Department of Mathematics from 1953 through 1963.

Interspersed throughout his service to Princeton were leaves and participation in other institutions. During World War II he taught in the Army Specialized Training Program and the Navy Pre-Radar Program, was Associate Director of the Applied Mathematics Group, National Defense Research Committee and worked with the Office of Scientific Research and Development. He was Visiting Professor at Stanford University in 1949-1950; Philips Visitor at Haverford in 1953-1954 and 1958-1959; Fulbright Lecturer in Australia, 1956; Visiting Lecturer for the Mathematical Association in 1956-1957; guest lecturer at various European universities for the Organization of European Economic Cooperation in 1959; guest lecturer at the Rockefeller Institute in 1963-1965; Visiting Professor at Darmouth in 1963; Visiting Professor at Arizona State University in 1971; and consultant to or member of many organizations such as the Rand Corporation, IBM Research, the Alfred P. Sloan Foundation, the President's Committee on the National Medal of Science, and the President's Science Advisory Committee. In 1974, in "retirement", Al Tucker becomes Mary Shepard Upson Visiting Professor of Engineering at Cornell.

In parallel with these academic duties Al Tucker has led an active existence as a leader of the mathematical community. His consistently wise counsel and ability to get things done led to his election as Council Member and Trustee of the American Mathematical Society, President of the Mathematical Association of America, a Vice President of the American Association for the Advancement of Science, and (first)

Chairman of the Conference Board of the Mathematical Sciences. In addition, he has been in the vanguard of efforts to update the mathematical curriculum, participating in a panoply of joints efforts at virtually all educational levels: the Committee (and Commission) on the Undergraduate Program in Mathematics, Chairman of the Commission on Mathematics of the CEEB (College Entrance Examination Board); the School Mathematics Study Group; Chairman of the Committee on Advance Placement of the CEEB; advisor to the Secondary School Mathematics Curriculum Improvement Study; lecturer at NSF Summer Institutes for teachers. For all this—and much more—Al Tucker was honored with the Award for Distinguished Service to Mathematics by the Mathematical Association of America in 1968.

Al Tucker has also been a strong influence on mathematical publications. He organized and managed the *Princeton Mathematical Notes, 1934-1939,* and their successor the *Annals of Mathematics Studies* from 1941 to 1949. He was founding editor of the *Princeton Mathematical Series* and a member of the editorial boards of the *Journal of Combinatorial Theory* and of *Mathematical Programming* since their beginnings.

This is but a recitation of his posts and works and awards, and although it provides rich evidence of his contributions, it misses the flavor of his particular mark on many individuals. One need only recall the many prefaces—indeed dedications—found at the beginnings of books; learn, first hand from friends, of the profound influence Al Tucker had upon their careers; know of the many who seek his advice on matters of importance to them. His research contributions are evidenced by his papers, but much more substantially, through the work of his students and colleagues, stimulated and encouraged by him. He has posed the right questions. He has maintained the perfectionist's reticence from writing too fast or too much. In a word, Al Tucker is and has been that rare person : a real teacher.

It is surely impossible to improve upon John Sloan Dickey's charge to Al Tucker upon presentation of the honorary degree of Doctor of Science in June 1961 at Darmouth College :

"Nearly three decades ago you began an academic career at Princeton which became a mission to mathematics. In a field where scholarship scores only if the idea is both new and demonstrably true your ideas have won their way in topology, in the theory of games, and in linear programming. But even in mathematics a mission is more than ideas; it is also always a man, a man who cares to the point of dedication, whose concern is that others should care too, and who can minister to the other fellow, as the need may be, either help or forbearance. Because you, sir, embody in extraordinary measure both your profession's love of precision and man's need for conscientious leadership, mathematics in America at all levels is today higher than it was and tomorrow will be higher."

M.L. Balinski
22 July 1974

Mathematical Programming Study 1 (1974) 4–9. North-Holland Publishing Company

BIBLIOGRAPHY*

of Albert William Tucker

to July 1974

"Generalised covariant differentiation", *Annals of Mathematics* 32 (1931) 451-460.

"On combinatorial topology", *Proceeding of the National Academy of Sciences USA* 18 (1932) 86-89.

"Modular homology characters", *Proceedings of the National Academy of Sciences USA* 18 (1932) 467-471.

"An abstract approach to manifolds" (Ph.D. Thesis), *Annals of Mathematics* 34 (1933) 191-243.

"Tensor invariance in the calculus of variations", *Annals of Mathematics* 35 (1934) 341-350.

"Non-Riemannian subspaces", *Annals of Mathematics* 36 (1935) 965-983.

"Two books on topology", *Bulletin of the American Mathematical Society* 41 (1935) 468-471.

"The topological congress in Moscow", *Bulletin of the American Mathematical Society* 41 (1935) 764.

"Cell-spaces", *Annals of Mathematics* 37 (1936) 92-100.

"Cell-spaces", *Matematičeskiĭ Sbornik* n.s. 1 (1936) 773-774.

"Alexandroff and Hopf on topology", *Bulletin of the American Mathematical Society* 42 (1936) 782-784.

"Branched and folded coverings", *Bulletin of the American Mathematical Society* 42 (1936) 859-862.

"Application of tensors to electrical engineering", *Electrical Engineering* 56 (1937) 619-620.

"Degenerate cycles bound", *Matematičeskiĭ Sbornik* n.s. 3 (1938) 287-288.

* Particularly significant reports, reviews, abstracts and public addresses have been included along with published papers.

"Chain mappings carried by cell-mappings", *Proceedings of the National Academy of Sciences USA* 25 (1939) 371-374.

"Algebraic structure of complexes", *Proceedings of the National Academy of Sciences USA* 25 (1939) 643-647.

"A boundary-value theorem for harmonic tensors" (Abstract), *Bulletin of the American Mathematical Society* 47 (1941) 714. ["Boundary theorems of tensor analysis in the large", 1941 manuscript, unpublished (a war casualty).]

"Some topological properties of disk and sphere", in: *Proceedings of the First Canadian Mathematical Congress*, Montreal, 1945 (The University of Toronto Press, Toronto, 1946) pp. 285-309.

"Some theorems on the sphere", in: S. Lefschetz, *Introduction to Topology* (Princeton University Press, Princeton, N.J., 1949) pp. 134-141.

"Topology" (with H.S. Bailey, Jr.), *Scientific American* 182 (1950) 18-24. Reprinted in: M. Kline, ed., *Mathematics in the modern world* (W.H. Freeman, San Francisco, Calif., 1968) pp. 134-140.

Contributions to the Theory of Games, H.W. Kuhn and A.W. Tucker, eds., Annals of Mathematics Study 24 (Princeton University Press, Princeton, N.J., 1950).

"On symmetric games" (with D. Gale and H.W. Kuhn), *Ibid.*, pp. 81-87.

"Reductions of game matrices" (with D. Gale and H.W. Kuhn), *Ibid.*, pp. 89-96.

"Linear programming and the theory of games" (with D. Gale and H.W. Kuhn), in: T.C. Koopmans, ed., *Activity Analysis of Production and Allocation* (John Wiley and Sons, New York, 1951) pp. 317-329. Pre-abstract: *Econometrica* 18 (1950) 189-190.

"Nonlinear programming" (with H.W. Kuhn), in: J. Neyman, ed., *Proceedings of the Second Berkeley Symposium on Mathematical Statistics and Probability* (University of California Press, Berkeley, Calif., 1951) pp. 481-492. Reprinted in: P. Newman, ed., *Readings in Mathematical Economics*, Volume I (Value Theory), (The Johns Hopkins Press, Baltimore, Md., 1968) pp. 3-14.

"On Kirchhoff's laws, potential, Lagrange multipliers, etc.", NAML Report 52-17, Institute for Numerical Analysis, University of California, Los Angeles, Calif. (1951).

"Theorems of alternatives for pairs of matrices", in: L. Goldstein and A. Orden, eds., *Symposium on Linear Inequalities and Programming*

(Project SCOOP, Planning Research Division, Director of Management and Analysis Service, Comptroller, Headquarters USAF, 1952) pp. 180-181.

Contributions to the Theory of Games, Volume II, H.W. Kuhn and A.W. Tucker, eds., Annals of Mathematics Study 28 (Princeton University Press, Princeton, N.J., 1953).

Game Theory and Programming, National Science Foundation Summer Mathematics Institute notes (The Oklahoma Agricultural and Mechanical College, Stillwater, Okla., 1955).

"Linear inequalities and convex polyhedral sets", in : H.A. Antosiewicz, ed., *Proceeding of the Second Symposium on Linear Programming* (U.S. National Bureau of Standards, Washington, D.C., 1956) pp. 569-602.

Linear Inequalities and Related Systems, H.W. Kuhn and A.W. Tucker, eds., Annals of Mathematics Study 38 (Princeton University Press, Princeton, 1956). [Russian translation : Moscow, 1959, edited by L.V. Kantorovich.]

"Dual systems of homogeneous linear relations", *Ibid.*, pp. 3-18.

"Polyhedral convex cones" (with A.J. Goldman), *Ibid.*, pp. 19-40.

"Theory of linear programming" (with A.J. Goldman), *Ibid.*, pp. 53-97.

"Linear programming", *Industrial Quality Control* 12 (1956) 1-4.

"Games, theory of" (with H.W. Kuhn), in : *Encyclopaedia Britannica*, 1956 edition.

Algebraic Geometry and Topology, a symposium in honor of S. Lefschetz, edited by R.H. Fox, D.C. Spencer and A.W. Tucker (Princeton University Press, Princeton, N.J., 1957).

"Linear and nonlinear programming", *Operations Research* 5 (1957) 244-257.

Contributions to the Theory of Games, Volume III, M. Dresher, A.W. Tucker and P. Wolfe, eds. (Princeton University Press, Princeton, N.J., 1957).

"Mathematical programming", retiring vice-presidential address, American Association for the Advancement of Science (December 1957) unpublished.

"John von Neumann's work in the theory of games and mathematical economics" (with H.W. Kuhn), *Bulletin of the American Mathematical Society* 64 (1958) 100-122.

Program for College Preparatory Mathematics, Report of the Commission on Mathematics, A.W. Tucker, chairman (College Entrance Examination Board, New York, 1959).

"The education of mathematics teachers in analysis", *American Mathematical Monthly* 66 (1959) 808-809.

Contributions to the Theory of Games, Volume IV, A.W. Tucker and R.D. Luce, eds., Annals of Mathematics Study 40 (Princeton University Press, Princeton, N.J., 1959).

"A combinatorial equivalence of matrices", in : R. Bellman and M. Hall, Jr., eds., *Combinatorial Analysis*, Proceedings of Symposia in Applied Mathematics, Volume X (American Mathematical Society, Providence, R.I., 1960) pp. 129-140.

"Abstract structure of the simplex method", in: P. Wolfe, ed., *The Rand Symposium on Mathematical Programming,* Proceedings of a Conference, March 16-20, 1959 (The Rand Corporation, Santa Monica, Calif., 1960) pp. 33-34.

"Abstract structure of linear programming", in : *Information Processing*, Proceeding of the International Conference on Information Processing, Paris, June 15-20, 1959 (UNESCO, Paris, Munich, London, 1960) pp. 99-100.

"Integer programming formulation of the travelling salesman problem" (with C.E. Miller and R.A. Zemlin), *Association for Computing Machinery Journal* 1 (1960) 326-329.

"Solving a matrix game by linear programming", *IBM Journal* 4 (1960) 507-517.

"Applications of mathematics", in: H.F. Fehr, ed.,*New Thinking in School Mathematics* (Organisation for European Economic Cooperation, Office for Scientific and Technical Personnel, Paris, 1961) pp. 49-60.

"Combinatorial classes of fair games", in: *La Décision* (Colloques Internationaux du Centre National de la Recherche Scientifique, Paris, 1961).

"Combinatorial equivalence of fair games", in: M. Maschler, ed., *Recent Advances in Game Theory,* Princeton University Conference, Princeton, N.J. (1962) pp. 277-282.

"A linear-convex existence theorem", in : *Abstracts of Short Communications,* International Congress of Mathematicians, Stockholm (1962) p. 203.

"Pivotal methods in linear algebra", retiring presidential address, Mathematical Association of America (January 1963), unpublished.

"Combinatorial theory underlying linear programs", in : R.L. Graves and P. Wolfe, eds., *Recent Advances in Mathematical Programming* (McGraw-Hill, New York, 1963) pp. 1-16.

"Simplex method and theory", in : R. Bellman, ed., *Mathematical Optimization Techniques* (University of California Press, Berkeley and Los Angeles, Calif., 1963) pp. 213-231.

"Principal pivot transforms of square matrices", in : J.R. Edmonds, Jr., ed., *Graphs and Combinatorics Conference* (Department of Mathematics, Princeton University, 1963). Also *SIAM Review* 5 (1963) 305.

"Linear programming and combinatorial mathematics", in : J.G. Kemeny, R. Robinson and R.W. Ritchie, eds., *New Directions in Matics* (Prentice-Hall, Englewood Cliffs, N.J., 1963) pp. 77-91.

"Combinatorial algebra of matrix games and linear programs", in : E.F. Beckenbach, ed., *Applied Combinatorial Mathematics* (John Wiley and Sons, New York, 1964) pp. 320-347.

Advances in Game Theory, M. Dresher, L.S. Shapley and A.W. Tucker, eds., Annals of Mathematics Study 52 (Princeton University Press, Princeton, N.J., 1964).

Pivotal Algebra, seminar notes by T.D. Parsons, mimeographed, Department of Mathematics, Princeton University, Princeton, N.J. (1965).

"John von Neumann, 1903-1957" (with S. Ulam, H.W. Kuhn and C.E. Shannon), in : *Perspectives in American History,* Volume II (Charles Warren Center for Studies in American History, Harvard University, Cambridge, Mass., 1968) pp. 235-269.

"Complementary slackness in dual linear subspaces", in : G.B. Dantzig and A.F. Veinott, Jr., eds., *Mathematics of the Decision Sciences,* Part 1, Lectures in Applied Mathematics Volume 11 (American Mathematical Society, Providence, R.I., 1968) pp. 137-143.

"A least-distance approach to quadratic programming", *Ibid.,* pp. 163-176.

"Least distance programming", in : R. Fletcher, ed., *Optimization* (Academic Press, London and New York, 1969) pp. 271-272.

"Duality theory of linear programs : a constructive approach with applications" (with M.L. Balinski), *SIAM Review* 11 (1969) 347-377.

"Least-distance programming", in : H.W. Kuhn, ed., *Proceedings of the Princeton Symposium on Mathematical Programming* (Princeton University Press, Princeton, N.J., 1970) pp. 583-588.

"Reduced matrices of graphs" (with T.D. Parsons), in: *Proceedings of the Second Chapel Hill Conference on Combinatorial Mathematics and its Applications* (University of North Carolina, Chapel Hill, N.C., 1970) pp. 373-380.

"Hybrid programs: linear and least-distance" (with T.D. Parsons), *Mathematical Programming* 1 (1971) 153-167.

Constructive Linear Algebra (with A. Gewirtz and H. Sitomer) (Prentice-Hall, Englewood Cliffs, N.J., 1974).

Mathematical Programming Study 1 (1974) 10. North-Holland Publishing Company

IN HONOR OF A.W. TUCKER'S CONTRIBUTIONS TO MATHEMATICAL PROGRAMMING

George B. DANTZIG

Some people are honored because they have invented some famous theorem that bears their name—others are known because of their foresight, imagination and leadership at crucial times in history. Albert W. Tucker's contributions to mathematical programming are many, but most important, I feel, are his contributions as a great unifier, expositor, coordinator and pioneer, who from the beginning, saw its potentialities and who excited the younger generation to venture into this new and unknown field.

In the summer of 1949, a most remarkable conference under the auspices of the Cowles Foundation and RAND Corporation took place at the University of Chicago. Mathematicians, economists and statisticians from academic and govermental institutions presented there their research on linear programming. The problems considered ranged from scheduling of crop rotation to the planning of large enterprises.

In retrospect, what is most surprising is that this conference was a report about a vast surge of research that had taken place in just two short years, for mathematical programming as a discipline was non-existant before 1947.

What made possible this remarkable outpouring (which has continued unabated to this day) was the foresight, imagination and leadership of two men, the Economist T.J. Koopmans and the Mathematician A.W. Tucker. It was they who inspired some of the finest minds of our generation to join the ranks.

Princeton, from 1948 on, with Tucker as coordinator, became the center for mathematical development of game theory, linear and mathematical programming. It was he with his students David Gale and Harold Kuhn, who began the work of setting forth the standards for a rigorous, constructive mathematical theory.

We are deeply indebted to Al Tucker, as a coordinator who inspired the young, who saw the importance of developing a unified theory and who used his talents as an expositor to make linear programming and its extensions known around the globe.

Mathematical Programming Study 1 (1974) 11–19. North-Holland Publishing Company

LOWER BOUNDS FOR MAXIMUM DIAMETERS
OF POLYTOPES*

Ilan ADLER

University of California, Berkeley, Calif. U.S.A.

Received 17 May 1972
Revised manuscript received 24 April 1974

The maximum diameter over all d-dimensional polytopes with n facets, $\Delta(d,n)$, represents the number of iterations required to solve the "worst" linear program using the ideal vertex-following algorithm. Hence $\Delta(d,n)$ measures, in a sense, the theoretical efficiency of such algorithms.

The main result of the paper is that $\Delta(d,n) \geq [(n-d) - (n-d)/[5d/4]] + 1$ for $n \geq d + 1$, and that these bounds are sharp for all known values of $\Delta(d,n)$.

0. Introduction

The diameter of a given polytope P is defined as the smallest integer k such that any two vertices of P can be joined by a path (of adjacent vertices) of length less than or equal to k. Let us denote by $\Delta(d, n)$ the maximum diameter of all d-dimensional polytopes with n facets.

The main result of this paper is the presentation of improved lower bounds for $\Delta(d, n)$.

The investigation of maximum diameters of polytopes is closely related to the study of efficiency of "vertex following" algorithms of linear programming, which start with a vertex and proceeds along successive adjacent vertices, according to some specified rule, until an optimal vertex is reached. Since, the maximum diameter of d-dimensional polytopes with n facets represents, in a sense, the number of iterations required to solve the "worst" linear program with $n - m$ equations in n

* This research has been supported by the National Science Foundation under Grant GP-30961X with the University of California.

nonnegative variables using the "best" vertex following algorithm. The main tools which are used to establish the new lower bounds for $\Delta(d, n)$ is the construction of product and summation of simple polytopes, those constructions are introduced in Section 2 together with some preliminary theorems. Then in Section 3 we present and prove the main result of the paper, namely that

$$\Delta(d, n) \geq \left[(n - d) - \frac{(n - d)}{[5d/4]}\right] + 1 \qquad (n \geq d + 1),$$

and that these bounds are sharp for all the known values of $\Delta(d, n)$. It is also shown that these new bounds are slightly better than previously known lower bounds which were presented by Klee [4] and Klee and Walkup [5].

1. Notations and definitions

A *convex polytope* (or simply a *polytope*) is a bounded nonempty intersection of a finite number of closed half spaces in a finite-dimensional real vector space. The *faces* of a polytope P are the intersections of P with its various supporting hyperplanes. Zero-, one- and $(d - 1)$-dimensional faces of a d-dimensional polytope P are called, respectively, the *vertices*, *edges* and *facets* of P. Two faces are said to be *incident* if one contains the other. A d-dimensional polytope is *simple* if each of its vertices is incident to exactly d edges.

Since it was shown by Klee and Walkup [5] that $\Delta(d, n)$ can be determined by considering only *simple* polytopes, we shall restrict our attention to simple polytopes and shall denote by $\mathcal{P}(d, n)$ the set of all d-dimensional simple polytopes with n facets.

As usual, $[x]$ denotes the largest integer less than or equal to x.

2. Product and sum of polytopes

2.1. *Product of polytopes*

Let $P_i \in \mathcal{P}(d_i, n_i)$ $(i = 1,2)$. We define the *product* $P_1 \otimes P_2$ of P_1

and P_2 by

$$P_1 \otimes P_2 = \{(x_1, x_2): x_i \in P_i; i = 1, 2\}$$

Theorem 2.1.

$$P_1 \otimes P_2 \in \mathscr{P}(d_1 + d_2, n_1 + n_2).$$

Proof. The proof follows directly from the definition.

2.2. *Sum of simple polytopes*

The following construction was suggested by Barnett [3], and its combinatorial equivalent independently by Adler [1]. The following discussion follows the one given in [3].

Let $P_i \in \mathscr{P}(d, n_i)$, $i = 1, 2$.

(1) Choose arbitrarily two vertices v_1 and v_2 from P_1 and P_2, respectively.

(2) Truncate vertices v_i producing polytopes P_i^1 with simplical facets F_i ($i = 1, 2$) which were created by the truncation.

(3) Take a hyperplane H passing through v_1 and apply a projective transformation τ_1 which sends H to infinity. In $\tau_1(P_1^1)$, all facets meeting $\tau_1(F_1)$ will be parallel. Apply the same kind of transformation τ_2 to P_2^1.

(4) Apply an affine transformation α_1 to $\tau_1(P_1^1)$ which will produce a polytope $P_1^2 = \alpha_1[\tau_1(P_1^1)]$ in which one facet meeting $\alpha_1[\tau_1(F_1)]$ is perpendicular to it. Note that all facets meeting $\alpha_1[\tau_1(F_1)]$ will be perpendicular to it. Apply the same kind of affine transformation α_2 to $\tau_2(P_2^1)$ to produce $P_2^2 = \alpha_2[\tau_2(P_2^1)]$.

(5) Apply an affine transformation α_3 to P_1^2 which will take $\alpha_1[\tau_1(F_1)]$ onto $\alpha_2[\tau_2(F_2)]$ and leaves the faces meeting $\alpha_1[\tau_1(F_1)]$ perpendicular to it.

(6) Place P_2^2 and $\alpha_3(P_1^2)$ so that $\alpha_3[\alpha_1(\tau_1(F_1))]$ and $\alpha_2[\tau_2(F_2)]$ coincide and so that the interior of P_2^2 misses the interior of $\alpha_3(P_1^2)$.

The polytope produced by this process will be called the *sum* of P_1 and P_2 and be denoted by $P_1 \oplus P_2$. Note that $P_1 \oplus P_2$ depends on the choice of v_1 and v_2 together with the choice of the several transformation mentioned above. For simplicity, we omit this dependence from the notation.

Note that all the facets of P_i (after the transformation) which do not contain v_i $(i = 1, 2)$ are facets of $P_1 \oplus P_2$ and that the d facets of P_1 which intersect at v_1 together with the d facets of P_2 which intersect at v_2 form (after the transformations) the remaining d facets of $P_1 \oplus P_2$.

Theorem 2.2.

$$P_1 \oplus P_2 \in \mathscr{P}(d, n_1 + n_2 - d).$$

Proof. The proof follows immediately from the definition and the comment following it.

3. Lower bounds for maximum diameters of polytopes

Let P be a polytope and let $v, \bar{v}$ be vertices of P. A *path* of *length* k from v to $\bar{v}$ in P is a sequence of vertices $v = v_0, \ldots, v_k = \bar{v}$ such that v_i, v_{i+1} are neighbors $(i = 0, \ldots, k - 1)$. The *distance* $\rho_P(v, \bar{v})$ between v and $\bar{v}$ in P is the length of the shortest path joining v and $\bar{v}$ in P. The *diameter* $\delta(P)$ of P is defined by

$$\delta(P) = \max \{\rho_P(v, \bar{v}): v, \bar{v} \in P\}.$$

Let us define $\Delta(d, n)$ as the maximum of $\delta(P)$, where P ranges over all d-dimensional polytopes with n facets.

We shall use the following two theorems in the construction of the lower bounds for $\Delta(d, n)$.

Theorem 3.1. *Let* $P_i \in \mathscr{P}(d_i, n_i)$, $i = 1,2$. *Then*
  (i) $\delta(P_1 \otimes P_2) = \delta(P_1) + \delta(P_2)$.
  (ii) *If* $d_1 = d_2$, *then one can sum* P_1 *and* P_2 *such that*

$$\delta(P_1) + \delta(P_2) - 1 \le \delta(P_1 \oplus P_2) \le \delta(P_1) + \delta(P_2).$$

Proof. (i) Let (v_1, v_2), $(\bar{v}_1, \bar{v}_2)$ be vertices of $P_1 \otimes P_2$, where $v_i, \bar{v}_i$ are vertices of P_i $(i = 1, 2)$. Let $v_i = v_i^0, \ldots, v_i^{k_i} = \bar{v}_i$ be the shortest path from v_i to $\bar{v}_i$ in P_i $(i = 1, 2)$. Then

$$(v_1, v_2) = (v_1^0, v_2), \ldots, (v_1^{k_1}, v_2)$$
$$= (\bar{v}_1, v_2^0), \ldots, (\bar{v}_1, v_2^{k_2}) = (\bar{v}_1, \bar{v}_2)$$

is a path of length $k_1 + k_2$ joining (v_1, v_2) to $(\bar{v}_1, \bar{v}_2)$ in $P_1 \otimes P_2$. Hence,

$$\rho_{P_1 \otimes P_2}((v_1, v_2), (\bar{v}_1, \bar{v}_2)) \leq \rho_{P_1}(v_1, \bar{v}_1) + \rho_{P_2}(v_2, \bar{v}_2).$$

Furthermore, if (u_1, u_2), $(\bar{u}_1, \bar{u}_2)$ is a pair of adjacent vertices in $P_1 \otimes P_2$, where $u_i, \bar{u}_i \in P_i$ $(i = 1, 2)$, then either $u_1 = \bar{u}_1$ and u_2 is adjacent to $\bar{u}_2$ in P_2, or $u_2 = \bar{u}_2$ and u_1 is adjacent to $\bar{u}_1$ in P_1. Thus

$$\rho_{P_1 \otimes P_2}((v_1, v_2), (\bar{v}_1, \bar{v}_2)) \geq \rho_{P_1}(v_1, \bar{v}_1) + \rho_{P_2}(v_2, \bar{v}_2).$$

The last two inequalities imply that

$$\rho_{P_1 \otimes P_2}((v_1, v_2), (\bar{v}_1, \bar{v}_2)) = \rho_{P_1}(v_1, \bar{v}_1) + \rho_{P_2}(v_2, \bar{v}_2).$$

So $\delta(P_1 \otimes P_2) = \delta(P_1) + \delta(P_2)$.

(ii) Let $v_i, \bar{v}_i \in P_i$ such that $\rho_{P_i}(v_i, \bar{v}_i) = \delta(P_i)$, $i = 1, 2$. Now let us sum P_1 and P_2 taking v_1 and v_2 as the two vertices which are eliminated in the summation construction.

Since $\rho(v_i, \bar{v}_i) = \delta(P_i)$, it is obvious that if v_i' is adjacent to v_i in P_i, then $\rho_{P_i}(v_i', \bar{v}_i)$ is equal to either $\delta(P_i)$ or to $\delta(P_i) - 1$, $i = 1, 2$. But every vertex in P_1 which is a neighbor of v_1 has exactly one adjacent vertex in P_2 which is a neighbor of v_2 and no other vertex of P_1 has adjacent vertex in P_2. Hence,

$$\delta(P_1) + \delta(P_2) - 1 \leq \delta(P_1 \oplus P_2) \leq \delta(P_1) + \delta(P_2).$$

Theorem 3.2. (i) $\Delta(d_1 + d_2, n_1 + n_2) \geq \Delta(d_1, n_1) + \Delta(d_2, n_2)$ *and in particular,* $\Delta(d + 1, n + 2) \geq \Delta(d, n) + 1$.
(ii) $\Delta(d, n_1 + n_2 - d) \geq \Delta(d, n_1) + \Delta(d, n_2) - 1$.

Proof. (i) Let $P_i \in \mathcal{P}(d_i, n_i)$, where $\delta(P_i) = \Delta(d_i, n_i)$. By Theorem 2.1, $P_1 \otimes P_2 \in \mathcal{P}(d_1 + d_2, n_1 + n_2)$; hence by Theorem 3.1,

$$\Delta(d_1 + d_2, n_1 + n_2) \geq \delta(P_1 \otimes P_2) = \delta(P_1) + \delta(P_2)$$

$$= \Delta(d_1, n_1) + \Delta(d_2, n_2).$$

If we let $P_1 \in \mathcal{P}(1,2)$ (i.e., P_1 is constituted from two adjacent vertices) then, since $\Delta(1, 2) = 1$,

$$\Delta(d + 1, n + 2) \geq \Delta(d, n) + 1.$$

(ii) Let $P_i \in \mathcal{P}(d, n_i)$, where $\delta(P_i) = \Delta(d, n_i)$ $(i = 1,2)$. By Theorem 2.2, $P_1 \oplus P_2 \in \mathcal{P}(d, n_1 + n_2 - d)$; hence by Theorem 3.1 (summing P_1 and P_2 as specified in this theorem),

$$\Delta(d, n_1 + n_2 - d) \geq \delta(P_1 \oplus P_2) \geq \delta(P_1) + \delta(P_2) - 1$$
$$= \Delta(d, n_1) + \Delta(d, n_2) - 1.$$

We are ready now to introduce the lower bounds for $\Delta(d, n)$.

Theorem 3.3.

$$\Delta(d, n) \geq \left[(n - d) - \frac{(n - d)}{[5d/4]} \right] + 1 \quad (n \geq d + 1).$$

Proof. Let

$$Z(d, n) = \left[(n - d) - \frac{(n - d)}{[5d/4]} \right] + 1.$$

It was shown by Klee and Walkup [5] that $\Delta(d, n) \geq Z(d, n)$ for $d \leq 2$. Assume that $\Delta(d - 1, n) \geq Z(d - 1, n)$ for some $d - 1 \geq 2$ and all $n \geq d$. By Theorem 3.2 and the induction assumption,

$$\Delta(d, n) \geq \Delta(d - 1, n - 2) + 1 \geq Z(d - 1, n - 2) + 1$$
$$= \left[(n - d - 1) - \frac{(n - d - 1)}{[5(d - 1)/4]} \right] + 2.$$

Suppose $d \neq 0 \pmod 4$, (i.e., $d/4$ is not an integer), then

$$Z(d - 1, n - 2) + 1 = \left[(n - d) - \frac{(n - d) - 1}{[5d/4] - 1} \right] + 1.$$

Thus, since $n - d \geq 1$,

$$Z(d - 1, n - 2) + 1 \geq Z(d, n) \quad \text{for } n - d \leq [5d/4].$$

Therefore,

$$\Delta(d, n) \geq Z(d, n) \quad \text{for } n - d \leq [5d/4] \quad (\text{and } d \neq 0 \,(\text{mod } 4)).$$

If $d = 0 \,(\text{mod } 4)$, then

$$Z(d - 1, n - 2) + 1 = \left[(n - d) - \frac{(n - d) - 1}{[5d/4] - 2} \right] + 1$$

and similarly to the previous case,

$$\Delta(d, n) \geq Z(d, n) \quad \text{for } n - d \leq [5d/4] - 1$$

$$(\text{and } d = 0 \,(\text{mod } 4)).$$

Furthermore, since $d = 0$ (mod 4), by Theorem 3.2 and because $\Delta(4, 9) = 5$ (see Adler and Dantzig [2]),

$$\Delta\left(d, d + \frac{5d}{4}\right) = \Delta\left(\frac{d}{4} \cdot 4, \frac{d}{4} \cdot 9\right) \geq \frac{d}{4}\Delta(4, 9) = \frac{5d}{4} = Z\left(d, d + \frac{5d}{4}\right).$$

Hence,

$$\Delta(d, n) \geq Z(d, n) \quad \text{for } n - d \leq [5d/4]$$

(regardless of whether $d = 0$ (mod 4) or $d \neq 0$ (mod 4).

Assume now that $\Delta(d, n) \geq Z(d, n)$ for $n \leq n_0$ (for some $n_0 \geq d + [5d/4]$). Let $(n_0 - d) = b \,\text{mod}([5d/4])$ (i.e., $(n_0 - d) - b = k[5d/4]$ for some integer k, where $0 \leq b < [5d/4]$).

By Theorem 3.2 and the induction assumption,

$$\Delta(d, n_0 + 1) \geq \Delta(d, n_0 - b) + \Delta(d, b + 1 + d) - 1$$

$$\geq Z(d, n_0 - b) + Z(d, b + 1 + d) - 1$$

$$= \left[(n_0 - b - d) - \frac{(n_0 - b - d)}{[5d/4]} \right] + 1$$

$$+ \left[(b + 1 + d - d) - \frac{(b + 1 + d - d)}{[5d/4]} \right] + 1 - 1$$

$$= \left[k[5d/4] - \frac{k[5d/4]}{[5d/4]} \right] + \left[(b + 1) - \frac{(b + 1)}{[5d/4]} \right] + 1$$

$$= \left[(n_0 + 1 - d) - \frac{(n_0 + 1 - d)}{[5d/4]} \right] + 1$$

$$= Z(d, n_0 + 1).$$

Hence, $\Delta(d, n) \geq Z(d, n)$ for all d and n for which $\Delta(d, n)$ is defined.

Remarks

(1) The previous known lower bounds for $\Delta(d, n)$ (Klee [4]) were

$$(d - 1)[n/d] - d + 2.$$

It is easily seen that the new bounds presented in Theorem 4.1 are slightly better since

$$(d - 1)\left[\frac{n}{d}\right] - d + 2 \leq \left[(n - d) - \frac{n - d}{d} \right] + 1$$

$$\leq \left[(n - d) - \frac{n - d}{[5d/4]} \right] + 1.$$

In fact, Klee and Walkup [5] showed that $\Delta(4, 9) = 5$ while the old lower bound for $\Delta(4, d)$ is $(4 - 1)[9/4] - 4 + 2 = 4$. Based on this value for $\Delta(4, 9)$, Klee and Walkup [5] introduce a table of lower bounds for $\Delta(d, n)$ for $d \leq 12$ and $n \leq 24 + 2d$. It can be checked that the new lower bounds given in Theorem 4.1 are slightly better then those given in this table.

(2) The new lower bounds for $\Delta(d, n)$ are sharp for all known values of $\Delta(d, n)$ (i.e., for $d = 1, 2, 3$ and for all n and d such that $n - d \leq 5$, see [5]).

(3) Purely combinatorial proofs and discussion for the bounds established in Theorem 3.3 are given via the construction of *Abstract Polytopes* in [1] and [2].

References

[1] I. Adler, "The Euler characteristic of abstract polytopes", Tech. Rept. No. 71–11, Department of Operations Research, Stanford University, Stanford, Calif. (1971).

[2] I. Adler and G.B. Dantzig, "Maximum diameter of abstract polytopes", *Mathematical Programming Study* 1 (1974) 20–40.

[3] D. Barnette, "A simple 4-dimensional nonfacet", *Israel Journal of Mathematics* 7 (1969) 16–20.

[4] V. Klee, "Diameters of polyhedral graphs", *Canadian Journal of Mathematics* 16 (1964) 602–614.

[5] V. Klee and D.W. Walkup, "The d-step conjecture for polyhedra of dimension $d < 6$", *Acta Mathematica* 117 (1967) 53–77.

Mathematical Programming Study 1 (1974) 20–40. North-Holland Publishing Company

MAXIMUM DIAMETER OF ABSTRACT POLYTOPES*

Ilan ADLER** and George B. DANTZIG

Stanford University, Stanford, Calif., U.S.A.

Received 29 November 1971
Revised manuscript received 15 April 1974

A combinatorial structure called *abstract polytope* is introduced. It is shown that abstract polytopes are a subclass of pseudo-manifolds and include (combinatorially) simple convex polytopes as a special case.

The main objective is to determine the maximum diameter of abstract polytopes of dimension less than or equal to 5. Those results are relevant to the study of the efficiency of "vertex following" algorithms since the maximum diameter of d-dimensional polytopes with n facets represent, in a sense, the number of iterations required to solve the "worst" problem (with constraint set of d variables with n inequality constraints) using the "best" vertex following algorithm.

1. Introduction

Several algorithms of optimization problems with polytopes as constraint set are based on what might be called "vertex following" methods. These algorithms are based on the identification of a special class of feasible solutions (called vertices) and the determination of an adjacency relation among them. A vertex following algorithm starts with a vertex and proceeds along successive adjacent vertices, according to some specified rule, until an optimal vertex is reached (or until it is shown that no optimal vertex exists). The simplex algorithm of linear programming or Lemke's algorithm for the linear complementarity problem can serve as examples.

Related to the efficiency of such vertex following algorithms is the

* Research and reproduction of this report was partially supported by Office of Naval Research, Contract N-00014-67-A-0112-0011; U.S. Atomic Energy Commission, Contract AT[04-3]326 PA #18; National Science Foundation, Grant GP 25738.

** Presently, University of California, Berkeley, Calif., U.S.A.

study of maximum diameter of polytopes. Loosely speaking, the maximum diameter of a d-dimensional polytope with n facets represents the number of iterations required to solve the "worst" problem (whose constraint set has $n - d$ equations and n nonnegative variables) in using the "best" vertex following algorithm.

In this paper, we introduce a general (and convenient) framework for investigating the structure of the vertex set of a polytope and its adjacency relations. This framework is provided by a set of three axioms which define *abstract polytopes*.

Our main objective is to establish values and bounds for the maximum diameter of abstract polytopes of dimension less than or equal to 5. These results are similar to those obtained by Klee and Walkup [4] for ordinary polytopes. Our results, however, apply to a more general class of combinatorial structures and imply theirs as a special case.

In Section 2, we introduce the three axioms defining abstract polytopes together with some of the terminology to be used in the paper.

In Section 3, we discuss the relations between abstract and ordinary polytopes.

The fourth section is intended primarily for readers who are familiar with the combinatorial topology terminology. In this section, we study the close relations between abstract polytopes and pseudo-manifolds and provide a preview of our results in terms of the later.

In Section 5, we present some preliminary results which are used in the proofs of the key theorems of Section 6. Finally, in Section 7, we summarize our results with respect to maximum diameter of abstract polytopes.

2. Abstract polytope—definition and notation

Given a finite set T of symbols, a family P of subsets of T (called vertices) forms a *d-dimensional abstract polytope* if the following three axioms are satisfied:

(i) Every vertex of P has cardinality d.

(ii) Any subset of $d - 1$ symbols of T is either contained in no vertices of P or in exactly two (called *neighbors* or *adjacent*).

(iii) Given any pair of vertices $v, \bar{v} \in P$, there exists a sequence of vertices $v = v_0, \ldots, v_k = \bar{v}$ such that

(a) v_i, v_{i+1} are neighbors $(i = 0, \ldots, k - 1)$,

(b) $\{v \cap \bar{v}\} \subset v_i, (i = 0, \ldots, k)$.

It is convenient to delete from T all symbols that are not used to define vertices. Hence, we denote by $\bigcup P$ the set of all symbols which appear in at least one vertex (i.e., $\bigcup P = \{t : t \in v \text{ for some } v \in P\}$).

Let u be a subset of $\bigcup P$ such that $|u| = k$, ($|u|$ denotes the cardinality of u). If $P' = \{v \in P : v \supset u\}$ is nonempty, we say that P' is the *face* of P which is generated by u, and denote it by $F_P(u)$ or simply $F(u)$ if the abstract polytope P is clear. It is not difficult to verify that the family $\{v - u : v \in F_P(u)\}$ of subsets obtained by deleting u from each vertex of such a face is a $(d - k)$- dimensional abstract polytope. In the sequel, we shall use this property of faces extensively. Whenever we refer to the abstract polytope associated with a face, it is understood that the deleting of common symbols has been performed. Since $F_P(u)$ corresponds to a $(d - k)$-dimensional abstract polytope, we say that it is a $(d - k)$-dimensional face of P. Zero, one and $(d - 1)$-dimensional faces are called, respectively, *vertices*, *edges*, and *facets*. Let P have n facets.

A d-dimensional abstract polytope with n facets is called an (n, d)-*abstract polytope*. (Note that $n = |\bigcup P|$.) We denote by $\mathcal{P}(n, d)$ the class of all (n, d)-abstract polytopes.

The *graph* $G(P)$ of an abstract polytope P is the graph whose vertices and edges correspond 1–1 to the vertices and edges of P, respectively.

Let P be an abstract polytope and let $v, \bar{v} \in P$. A *path* of *length* k from v to $\bar{v}$ in P is a sequence of vertices $v = v_0, \dots, v_k = \bar{v}$ such that v_i, v_{i+1} are neighbors ($i = 0, \dots, k - 1$). (Note that vertices of the path are not required to be in $F_P(v \cap \bar{v})$.) The *distance* $\rho_P(v, \bar{v})$ between v and $\bar{v}$ in P is the length of the shortest path joining v and $\bar{v}$. The *diameter* $\delta(P)$ or δP is the smallest integer k such that any two vertices of P can be joined by a path of length less than or equal to k: $\delta(P) = \max \rho_P(v, \bar{v})$ for $v, \bar{v} \in P$. We denote by $\Delta_a(n, d)$ the maximum of $\delta(P)$ over all (n, d)-abstract polytopes. This corresponds to Klee and Walkup's $\Delta_b(n, d)$ for ordinary simple polytopes [4]. In general, of course, $\Delta_a(n, d) \geq \Delta_b(n, d)$.

As stated in the Introduction, our main objective is to establish values and bounds for $\Delta_a(n, d)$. We shall show in particular that the analog of the unsolved d-step (or Hirsch) conjecture, i.e., that $\Delta_a(n, d) = n - d$ holds for $n - d \leq 5$.

3. Relation between abstract and simple polytopes

Abstract polytopes are (combinatorially) closely related to simple polytopes. A simple polytope can be expressed as the set of solutions of a bounded and non-degenerate linear program. Suppose the latter consists of m equations in n non-negative variables whose coefficient matrix is of rank m. One can associate n symbols with the index set of the n columns of the coefficient matrix. Then the family of subsets of symbols which correspond to the non-basic columns of all the basic feasible solutions (i.e., vertices) of the linear program forms an (n, d)-abstract polytope where $d = n - m$. This is true because any feasible solution is defined uniquely by the subset of $d = n - m$ non-basic variables set to zero (axiom (i)). Given a basic feasible solution, a new basic solution can be obtained by dropping any one of the d non-basic variables. Exactly one of the basic variables can be set equal to zero in its place (under non-degeneracy and boundedness). This generates a neighboring vertex (axiom (ii)). Given any two vertices v and $\bar{v}$, then by restricting ourselves to the lowest dimensional face common to v and $\bar{v}$ (i.e., holding at zero value the subset of non-basic variables common to the two vertices), a path of neighboring vertices from v to $\bar{v}$ can be found (e.g., by using the simplex method and a suitably chosen objective function)(axiom (iii)).

Although the class of abstract polytopes includes (combinatorially) that of simple polytopes, the converse is not true. Indeed, it is well known (e.g., see [2, p. 235]) that the graph of 3-dimensional simple polytope is planar. However, the graph of the 3-dimensional abstract polytope displayed in Fig. 2 is easily shown to be non-planar. Hence no simple polytope can have the graph structure of this particular abstract polytope. See also Remark 6.10.

4. Abstract polytopes and pseudo-manifolds

In this section, we explore the very close association between abstract polytopes and pseudo-manifolds. This section is intended primarily for the readers familiar with combinatorial topology terminology to provide them with a "dictionary" relating our own terminology with that of pseudo-manifolds. Since the rest of the paper is self-contained, this section can be skipped.

Definition 4.1. A *simplicial complex K* consist of a set $\{v\}$ of vertices and a set $\{s\}$ of nonempty subsets of $\{v\}$ called *simplices* such that
   (i) any set consisting of exactly one vertex is a simplex,
   (ii) any nonempty subset of a simplex is a simplex.

Definition 4.2. The *dimension* of a simplex s containing $d + 1$ vertices is defined to be d and such a simplex is called a *d-simplex*. If $s' \subset s$, then s' is called a *face* of s and if s' is d'-simplex, then it is called a *d'-face* of s.

Definition 4.3. A *d-dimensional pseudo-manifold* without boundary (or simply a *d-pseudo-manifold*) is a simplicial complex K such that
    (i) every simplex of K is a face of some d-simplex of K,
    (ii) every $(d - 1)$-simplex of K is the face of exactly two d-simplices of K,
    (iii) if s and s' are d-simplices of K, there is a finite sequence $s = s_1, s_2, \ldots, s_k = s'$ of d-simplices of K such that s_i and s_{i+1} have a $(d - 1)$-face in common for $i = 1, \ldots, k - 1$.

Now, let us apply the following natural correspondence between abstract polytopes and pseudo-manifolds. Given a $(d + 1)$-dimensional abstract polytope P, one can associate with it a simplicial complex K as follows: Let $\bigcup P$ (see definition in Section 2) be the set of vertices of K and let $s \subset \bigcup P$ be a simplex of K if and only if $s \subset v$ for some $v \in P$. It is easily checked out that K is in fact a d-pseudo-manifold, since axioms (i)–(ii) are identical with the first two conditions of pseudo-manifolds and axiom (iii) is stronger than the third condition. However, if we try the reverse process of associating a $(d + 1)$-dimensional abstract polytope P to a given d-pseudo-manifold K (by defining the set of the vertices of P as the set of all d-simplices of K), we might fail because axiom (iii) is not necessarily satisfied. Therefore, let us restrict ourself to a special class of pseudo-manifolds as follows: Given a simplex s in a pseudo-manifold K, the *link of s in K* is the complex composed of all simplices of K which have no vertex in common with s, but which are faces of a simplex having s as a face. We say that a d-pseudo-manifold K is *locally connected*[1] if $d = 1$ or $d \geq 2$ and for each k-simplex s of K ($k \leq d - 2$) the link of s in K is a pseudo-manifold.

With a little patience, one can verify that the correspondence (defined above) between $(d + 1)$-dimensional abstract polytopes and d-dimensional locally connected pseudo-manifolds (LCPM) is one-to-one. Moreover, if $s' \subset \bigcup P$ generates an $(i + 1)$-dimensional face F of the abstract polytope P, then the link of s' in K (the corresponding LCPM) is the i-dimensional LCPM corresponding to F. Thus, abstract polytopes and

---

[1] This term was suggested by D. Walkup (private communication).

locally connected pseudo-manifolds are essentially identical combinatorial structures and one can use either of them in developing the results presented in this paper. Although the terminology of pseudo-manifolds is well established and widely used, we prefer the abstract polytope terminology because of its natural association to simple polytopes (or equivalently nondegenerate linear programs) which are our primary subject of investigation. However, we shall outline here our main results and line of proofs using the combinatorial topology terminology to provide a link between that terminology and ours.

In terms of pseudo-manifolds, our purpose is to find the maximum diameter of locally connected d-pseudo-manifolds with n vertices for $n - d + 1 \leq 5$. (The *diameter* of a pseudo-manifold K is defined as the smallest integer k such that any two simplices of K can be joined by a path of adjacent simplices of length less than or equal to k.) This result follows a similar one obtained by Klee and Walkup [4] for convex polytopes. Their main argument is that every simple 3-polytope with 6, 7 or 8 facets satisfies what they term as Property A (see [4] and Remark 6.3). But this line of proof cannot be used in our case because of the existence of a (unique) 2-dimensional locally connected pseudo-manifold with 8 vertices (see Fig. 2) which violates Property A. (In fact, this 2-pseudo-manifold corresponds to a triangulated projective plane with a handle.) However, considering the uniqueness of that counterexample and using some simple arguments, we can still prove the main result, namely, that the maximum diameter of a locally connected 4-pseudo-manifold with 10 vertices is 5. (See Theorems 6.1 and 7.1.)

5. Some preliminary results

We shall make frequent use of the following theorem.

Theorem 5.1 [1]. *Given an abstract polytope P, if two vertices v, $\bar{v}$ in P do not have a symbol (say A) in common, then there exists an "A-avoiding path" joining them; i.e., there exists a path from v to $\bar{v}$ such that no vertex along the path contains A.*

The next theorem is the analog of a result of Klee and Walkup [4]. The proof here is similar.

Theorem 5.2. *For* $k = 0, 1, 2, \ldots$
   (i) $\Delta_a(n, d) \leq \Delta_a(n + k, d + k)$,
   (ii) $\Delta_a(n, d) \leq \Delta_a(n + k, d)$,
   (iii) $\Delta_a(n, d) \leq \Delta_a(n + 2k, d + k) - k$, $\Delta_a(2d, d) \geq d$,
   (iv) $\Delta_a(2d, d) = \Delta_a(2d + k, d + k)$.

Proof. We shall prove (i)–(iii) for $k = 1$; the extension to $k > 1$ is trivial.
   Let P be an (n, d)-abstract polytope such that $\delta(P) = \Delta_a(n, d)$.
   (i) Let $A \in \bigcup P$ and let $A' \notin \bigcup P$ be a new symbol, define P' as an abstract polytope identical with P *except* the symbol A' replaces A. Define $\tilde{P}$ as a new abstract polytope with vertices $v \cup A'$ and $v' \cup A$ for all $v \in P$ and all $v' \in P'$.
   It is easy to verify that $\tilde{P}$ is an $(n + 1, d + 1)$-abstract polytope with a diameter at least as big as $\delta(P)$, thus

$$\Delta_a(n + 1, d + 1) \geq \delta(\tilde{P}) \geq \delta(P) = \Delta_a(n, d).$$

This inequality is sharp since it will be shown later that $\Delta_a(6, 2) = \Delta_a(7, 3) = 3$ and $\Delta_a(2d, d) = \Delta_a(2d + k, d + k)$ for all $k \geq 0$.
   (ii) Let $A' \notin \bigcup P$ be a new symbol and $v' \in P$. Let $v_1, \ldots, v_d$ be the d subsets of v' with cardinality $d - 1$. Define

$$\tilde{P} = P \setminus \{v'\} \cup \{v_i^* : v_i^* = v_i \cup A'; i = 1, \ldots, d\}.$$

It is obvious that $\tilde{P} \in \mathscr{P}(n + 1, d)$(i.e., $\tilde{P}$ is an $(n + 1, d)$-abstract polytope) and that $\delta(\tilde{P}) \geq \delta(P)$, hence

$$\Delta_a(n + 1, d) \geq \delta(\tilde{P}) \geq \delta(P) = \Delta_a(n, d).$$

This inequality is also sharp since it will be shown that $\Delta_a(n, 2) = \Delta_a(n + 1, 2) = n/2$ for n even.
   (iii) Let $A_1', A_2' \notin \bigcup P$ be two new distinct symbols. Define $P_i = \{(v \cup A_i') : v \in P\}$, $i = 1,2$. Then $P_1 \cup P_2 \in \mathscr{P}(n + 2, d + 1)$ and $\delta(P_1 \cup P_2) = \delta(P) + 1$. So

$$\Delta_a(n + 2, d + 1) - 1 \geq \delta(P_1 \cup P_2) - 1 = \delta(P) = \Delta_a(n, d).$$

In particular, $1 = \Delta_a(2, 1) \leq \Delta_a(2d, d) - (d - 1)$ or $\Delta_a(2d, d) \geq d$.
   (iv) Let $P \in \mathscr{P}(2d + k, d + k), (k \geq 0)$, where $\delta(P) = \Delta_a(2d + k, d + k)$.

Choose $v, \bar{v} \in P$ so that the shortest path from v to $\bar{v}$ has length $\delta(P)$. Note that $|v \cap \bar{v}| = k + \ell, 0 \leq \ell \leq d$. Consider the face $P' = F_P(v \cap \bar{v})$ of P which corresponds $1-1$ to a $(d + k' - j, k')$-abstract polytope, where $k' = d + k - |v \mathscr{P} \bar{v}| = d - \ell$ and $0 \leq j \leq k$. Since $P' \subset P$, the length of the shortest path from v to $\bar{v}$ in P' is at least as large as $\delta(P)$. Hence

$$\Delta_a(2d - \ell - j, d - \ell) \geq \delta(P') \geq \delta(P) = \Delta_a(2d + k, d + k).$$

However, by (i) and (ii),

$$\Delta_a(2d + k, d + k) \geq \Delta_a(2d, d) \geq \Delta_a(2d - \ell - j, d - \ell).$$

Hence

$$\Delta_a(2d + k, d + k) = \Delta_a(2d, d)$$

Theorem 5.3. *Given $d > 1$ and $k \geq 0$, there exists a $P \in \mathscr{P}(2d + k, d)$ with disjoint vertices $v^*, \bar{v}^*$ whose distance $\rho(v^*, \bar{v}^*)$ is $\Delta_a(2d + k, d)$.*

Proof. Let $P_1 \in \mathscr{P}(n, d)$, where $n \geq 2d$ have two vertices $v_0, \bar{v}_0$ such that $\rho_{P_1}(v_0, \bar{v}_0) = \Delta_a(n, d)$. Assume $s = |v_0 \cap \bar{v}_0| > 0$ and consider the face $P_2 = F_{P_1}(v_0 \cap \bar{v}_0) \in \mathscr{P}(n - s - j, d - s)$, where $j \geq 0$ is the number of symbols in the set $\bigcup P_1 \backslash \{v_0 \cap \bar{v}_0\}$ not used to form the vertices of the face. We have

$$\Delta_a(n, d) = \delta(P_1) \leq \delta F_{P_1}(v_0 \cap \bar{v}_0) \leq \Delta_a(n - s - j, d - s).$$

But by Theorem 5.2(i),(iii), $\Delta(n, d) \geq \Delta(n - s - j, d - s)$ with strict inequality if $j > 0$. Hence we conclude $j = 0$, $P_2 \in \mathscr{P}(n - s, d - s)$, and

$$\delta P_2 = \Delta_a(n - s, d - s) = \Delta_a(n, d).$$

Moreover, the vertices $v'_0 = v_0 \backslash \{v_0 \cap \bar{v}_0\}$, $\bar{v}'_0 = \bar{v}_0 \backslash \{v_0 \cap \bar{v}_0\}$ of P_2 are disjoint and

$$\rho_{P_2}(v'_0, \bar{v}'_0) = \Delta_a(n - s, d - s) = \Delta_a(n, d).$$

Let $u = \bigcup P_2 \backslash \{v'_0 \cup \bar{v}'_0\}$. Note $|u| = (n - s) - 2(d - s)$. Note also that $|u| \geq s$ if and only if $n \geq 2d$, i.e., if $n = 2d + k$, $k \geq 0$, which is in accord with our hypothesis. Select any subset of s symbols $\{A_1, A_2, \ldots, A_s\} \subset u$. We now define a new abstract polytope P by adjoining to $\bigcup P_2$ the set of new symbols $u' = \{A_1', A_2', \ldots, A_s'\}$. For each $v \in P_2$ we define $\{v \cup u'\} \in P$; given any generated $v' \in P$ and any i then replacing in v', A_i' by A_i if $A_i' \in v'$ and $A_i \notin v'$ generates additional $v'' \in P$. It is not difficult to show that $P \in \mathscr{P}(n, d)$ and $\delta P_2 \leq \delta P$; moreover, $\rho_{P_2}(v_0', \bar{v}_0') \leq \rho_P(v_0^*, \bar{v}_0^*)$, where $v_0^* = \{v_0' \cup u\}$ and $\bar{v}_0^* = \{\bar{v}_0' \cup u'\}$ Note that $v_0' \cap u' = \emptyset$, $\bar{v}_0' \cap u = \emptyset$ so that $v_0' \cap \bar{v}_0' = \emptyset$ implies that v_0^* and $\bar{v}_0^*$ are also disjoint. We have

$$\Delta_a(n, d) = \Delta_a(n - s, d - s) = \rho_{P_2}(v_0', \bar{v}_0') \leq \rho_P(v_0^*, \bar{v}_0^*) \leq \Delta_a(n, d).$$

Corollary 5.4. *There exists a $P \in \mathscr{P}(2d - k, d)$ with vertices v^* and $\bar{v}^*$ such that $\{\bigcup P - v^*\}$ and $\{\bigcup P - \bar{v}^*\}$ are disjoint and $\rho(v^*, \bar{v}^*) = \Delta_a(2d - k, d)$.*

Proof. The proof is along similar lines and will be omitted. Theorem 5.3 and Corollary 5.4 are the same for $k = 0$.

6. Key theorems

Theorem 6.1 will be used (in Section 7) to establish the values of $\Delta_a(n, 2)$ and the values of $\Delta_a(n, d)$ for all n, d such that $n - d \leq 5$. Note that Theorem 5.3 allows us to consider $P \in \mathscr{P}(2d, d)$, with disjoint vertex pairs $v_0, \bar{v}_0 \in P$ such that $\rho(v_0, \bar{v}_0) = \delta P = \Delta_a(2d, d)$.

We shall make frequent use of the notion of the "shell" bordering a set of vertices of an abstract polytope. Let P be an abstract polytope and let $Z \subset P$. A vertex v of P belongs to the ith *shell* $N_P^i(Z)$ of Z in P if and only if i is the minimum length of all the paths in P joining v to the various vertices of Z. The 0-shell of Z is Z itself. The 1-shell of Z is the set of vertices which are adjacent to but not in Z. In general,

$$N_P^i(Z) = N_P^1\left(\bigcup_{j=0}^{i-1} N_P^j(Z)\right).$$

For simplicity, the 1-shell of Z in P will also be denoted by $N_P(Z)$ or simply $N(Z)$ if P is clear.

Theorem 6.1. *Given* $P \in \mathcal{P}(2d, d)$ *(i.e., P is a (2d, d)-abstract polytope) and* $v_0, \bar{v}_0 \in P$ *such that* $v_0, \bar{v}_0$ *partition* $\bigcup P$. *Let* $(v_0, v_1, \ldots, v_k)$ *and* $(\bar{v}_0, \bar{v}_1, \ldots, \bar{v}_{\bar{k}})$ *be two paths in P with the property* $|v_i \cap \bar{v}_j| = i + j$, *then such paths exist for*

(i) $d \geq 1, k = 0, \bar{k} = 1$,
(ii) $d \geq 2, k = 1, \bar{k} = 1$,
(iii) $d \geq 3, k = 2, \bar{k} = 1$,

where $\bar{v}_1$ *is any given vertex in* $N(\bar{v}_0)$,

(iv) $d \geq 4, k = 2, \bar{k} = 2$.

Proof (except part (iv) for $d \geq 5$). It is convenient to switch from using symbols $A_i \in T$ to symbols $\{1, \ldots, d; \bar{1}, \ldots, \bar{d}\}$, and let $v_0 = \{1, \ldots, d\}$, $\bar{v}_0 = \{\bar{1}, \ldots, \bar{d}\}$ partition P. The symbols $v_i, \bar{v}_j$ where used below satisfy $|v_i \cap \bar{v}_j| = i + j$ or will be shown to do so.

(i) Obvious.

(ii) Relable so that $\bar{v}_1 = \{1, \bar{1}, \ldots, \overline{d-1}\}$. Note that $\{\bar{d}\} \not\subset (v_0 \cup \bar{v}_1)$. By Theorem 5.1, there exists an $\bar{d}$-avoiding path between v_0 and $\bar{v}_1$ in $F(v_0 \cap \bar{v}_1)$. Thus, all the vertices of the path contain $v_0 \cap \bar{v}_1 = \{1\}$ but do not contain $\{\bar{d}\}$. Let v_1 be the neighbor of v_0 in this path. Since $\{\bar{d}\} \not\subset v_1$, it must contain, for $d \geq 2$, one symbol different from those in $v_0 \cup \{\bar{d}\}$. Hence, noting $\{1\} \subset v_1, |v_1 \cap \bar{v}_1| = 2$.

(iii) Let $\bar{v}_1 \in N(\bar{v}_0)$, then by (ii) there exists a vertex $v_1 \in N(v_0)$ such that $|v_1 \cap \bar{v}_1| = 2$. By relabeling, let $v_1 = \{1, \ldots, d-1, \bar{1}\}$, $\bar{v}_1 = \{1, \bar{1}, \ldots, \overline{d-1}\}$. Define $P' = F(v_1 \cap \bar{v}_1)$ and $W = N(v_0) \cap P'$. Note that W is the set of all vertices of $N(v_0)$ which contain both $\{1\}$ and $\{\bar{1}\}$.

By Theorem 5.1, there exists a $\bar{d}$-avoiding path from v_1 to $\bar{v}_1$ in P'. Let v_2 be a vertex of this path which belongs to $N_{P'}(W)$ (such vertex exists because $v_1 \in W$ while $\bar{v}_1 \notin W$ for $d \geq 3$). But v_2 contains $\{1, \bar{1}\}$ and one symbol out of $\{\bar{2}, \ldots, \overline{d-1}\}$, hence $|v_2 \cap \bar{v}_1| = 3$.

(iv) ($d = 4$) by (iii), there exists $\bar{v}_1 \in N(\bar{v}_0)$ and $v_2 \in N^2(v_0)$ such that $|v_2 \cap \bar{v}_1| = 3$. Since $d = 4$, the second axiom of abstract polytopes implies that v_2 is a neighbor of $\bar{v}_1$. Thus letting $\bar{v}_2 = v_2$ completes the proof of this case.

Part (iv) for $d = 5$ will be established via Theorem 6.2 and for $d \geq 6$ via Theorems 6.4, 6.5, 6.7 and 6.9.

Theorem 6.2. *Let* $P \in \mathcal{P}(10, 5)$ *and let* $(v_0, v_1, v_2); (\bar{v}_0, \bar{v}_1)$ *be paths in P satisfying* $|v_i \cap \bar{v}_j| = i + j$. *Define*

$$P' = F(v_1 \cap \bar{v}_1), \qquad W = N(v_0) \cap P', \qquad \bar{W} = N(\bar{v}_0) \cap P'.$$

Then either there exists a path of length 3 connecting a vertex in W to a vertex in $\bar{W}$ or

(a) $F(v_2 \cap \bar{v}_1)$ *is (by relabeling) the 7-vertex 2-dimensional abstract polytope displayed in Fig. 1 and also by heavy edges in Fig. 2.*

(b) $F(v_1 \cap \bar{v}_1)$ *is (by relabeling) the 3-dimensional abstract polytope given in Fig. 2 (note that the graph of $F(v_1 \cap \bar{v}_1)$ is non-planar).*

$$v_2 = \left(\begin{array}{c} \{1,\bar{1},\bar{2}\} \cup \{2,3\}\text{---}\{1,\bar{1},\bar{2}\} \cup \{5,3\}\text{---}\{1,\bar{1},\bar{2}\} \cup \{5,\bar{4}\} \\[2mm] \{1,\bar{1},\bar{2}\} \cup \{2,\bar{5}\}\text{---}\{1,\bar{1},\bar{2}\} \cup \{4,\bar{5}\}\text{---}\{1,\bar{1},\bar{2}\} \cup \{4,\bar{3}\} \end{array}\right\rangle \{1,\bar{1},\bar{2}\} \cup \{3,\bar{4}\} = \bar{v}_1$$

Fig. 1. Face $F(1,\bar{1},\bar{2})$.

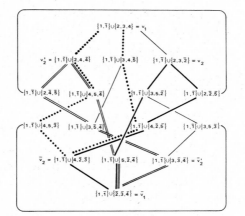

Fig. 2. Face $F(1, \bar{1})$. $F(1, \bar{1}) \in \mathscr{P}(8, 3)$ has a diameter $\delta = 4$.

Proof. Assume (by relabeling if necessary) that

$$v_0 = \{1, 2, 3, 4, 5\}, \qquad v_1 = \{1, 2, 3, 4, \bar{1}\}, \qquad v_2 = \{1, 2, 3, \bar{1}, \bar{2}\},$$
$$\bar{v}_0 = \{\bar{1}, \bar{2}, \bar{3}, \bar{4}, 5\}, \qquad \bar{v}_1 = \{\bar{1}, \bar{2}, \bar{3}, \bar{4}, 1\}.$$

(a) Since $|v_2 \cap \bar{v}_1| = 3$, $P'' = F(v_2 \cap \bar{v})$ corresponds 1–1 to an $(n, 2)$-abstract polytope. It is easy to show that every $(n, 2)$-abstract polytope

Q has exactly n vertices and that every symbol of $\bigcup Q$ is contained by two adjacent vertices of Q. Furthermore, the graph of Q forms a simple cycle with diameter $[n/2]$. It is obvious that the number of vertices in P'' satisfies

$$7 = \left| \bigcup P \right| - \left| v_2 \cap \bar{v}_1 \right| \geq \left| P'' \right| \geq \left| v_2 \cup \bar{v}_1 \right| - \left| v_2 \cap \bar{v}_1 \right| = 4.$$

(a1) $|P''| \leq 5$. In this case $n = 4$ (or 5), there exists a path of length two joining v_2 and $\bar{v}_1$, hence of length 3 joining v_1 in W to $\bar{v}_1$ in $\bar{W}$.

(a2) $|P''| = 6$. In this case P'' has the form

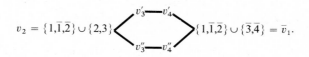

$$v_2 = \{1,\bar{1},\bar{2}\} \cup \{2,3\} \quad\quad \{1,\bar{1},\bar{2}\} \cup \{\bar{3},\bar{4}\} = \bar{v}_1.$$

Since $\{1, \bar{1}, \bar{2}\}$ is a subset of each vertex of P'', and $v_2 \cup \bar{v}_1 = \{1, \bar{1}, \bar{2}\} \cup \{2, 3, \bar{3}, \bar{4}\}$, one of the remaining symbols $\{4\}$, $\{5\}$, or $\{\bar{5}\}$ is contained by the two adjacent vertices v_3', v_4' and another by v_3'', v_4''. But if $\{4\}$ is contained by v_3', v_4' (or v_3'', v_4''), then v_3' (or v_3'') is a neighbor of v_1. If not, then $\{\bar{5}\}$ is contained by v_3', v_4' (or v_3'', v_4''), and v_4' (or v_4'') is a member of $\bar{W}$. In either case, there exists a path of length 3 from a member of W to a member of $\bar{W}$.

(a3) $|P''| = 7$. Here P'' has the form

$$v_2 = \{1,\bar{1},\bar{2}\} \cup \{2,3\} \quad\quad \{1,\bar{1},\bar{2}\} \cup \{\bar{3},\bar{4}\} = \bar{v}_1.$$

Using the same arguments as in (a2), we see that if every path from a member of W to a member of $\bar{W}$ is *not* to have a length of 3, then v_3', v_4' must contain $\{\bar{5}\}$; v_4', v_5' must contain $\{4\}$ and v_3'', v_4'' must contain $\{5\}$. Thus P'' has the form of Fig. 1, except for possible interchange of symbols $\{2\}$ with $\{3\}$ and $\{\bar{3}\}$ with $\{\bar{4}\}$.

(b) Suppose every path in P' joining a member of W to a member of $\bar{W}$ has a length larger than 3, so that P'' as displayed in Fig. 2 by heavy lines has the form given in Fig. 1. Let us denote the vertex $\{1, \bar{1}, \bar{2}\} \cup \{4, \bar{3}\} \in \{P''\}$ by $\bar{v}_2$. We can apply now the above analysis to the face $F(v_1 \cap \bar{v}_2)$, where we permute the symbols $\{1, 2, 3, 4, 5\}$ into $\{\bar{1}, \bar{3}, \bar{2}, \bar{4}, 5\}$

and $\{\bar{1}, \bar{2}, \bar{3}, \bar{4}, \bar{5}\}$ into $\{1, 4, 2, 3, 5\}$. Thus $F(v_1 \cap \bar{v}_2) = F(1, 4, \bar{1})$ has the following form (with the possible interchange of $\{2\}$ with $\{3\}$ and $\{\bar{2}\}$ with $\{\bar{3}\}$):

$$v_1 = \{1,4,\bar{1}\} \cup \{2,3\} \left\langle \begin{array}{l} \{1,4,\bar{1}\} \cup \{3,\bar{5}\} \!-\!\!\!-\! \{1,4,\bar{1}\} \cup \{2,\bar{5}\} \!-\!\!\!-\! \{1,4,\bar{1}\} \cup \{\bar{2},3\} \\[8pt] \{1,4,\bar{1}\} \cup \{2,\bar{4}\} \!-\!\!\!-\! \{1,4,\bar{1}\} \cup \{5,\bar{4}\} \!-\!\!\!-\! \{1,4,\bar{1}\} \cup \{5,\bar{3}\} \end{array} \right) = \bar{v}_2.$$

Note that interchanging $\{\bar{2}\}$ with $\{\bar{3}\}$ is not possible since $\{1, 4, \bar{1}\} \cup \{\bar{2}, \bar{5}\}$ is forced as a neighbor of $\bar{v}_2$ because it already exists as a vertex of $F(v_2 \cap \bar{v}_1)$. The above cycle is shown by dotted edges in Fig. 2 with the possibility that the symbol $\{2\}$ is interchanged with $\{3\}$ in two of its vertices.

We now let $v'_2 = \{1, \bar{1}, \bar{4}\} \cup \{2, 4\}$ and apply the argument of (a) on $F(v'_2 \cap \bar{v}_1) = F(1, \bar{1}, \bar{4})$. Since $\{1, \bar{1}, \bar{4}\} \cup \{5, \bar{2}\} \in F(v_2 \cap \bar{v}_1)$ and $\{1, \bar{1}, \bar{4}\} \cup \{4, 5\} \in F(v_1 \cap \bar{v}_2)$, we get a unique form for $F(v'_2 \cap \bar{v}_1)$ as follows:

$$v'_2 = \left(\begin{array}{l} \{1,\bar{1},\bar{4}\} \cup \{2,4\} \!-\!\!\!-\! \{1,\bar{1},\bar{4}\} \cup \{4,5\} \!-\!\!\!-\! \{1,\bar{1},\bar{4}\} \cup \{5,\bar{2}\} \\[8pt] \{1,\bar{1},\bar{4}\} \cup \{2,\bar{5}\} \!-\!\!\!-\! \{1,\bar{1},\bar{4}\} \cup \{3,\bar{5}\} \!-\!\!\!-\! \{1,\bar{1},\bar{4}\} \cup \{3,\bar{3}\} \end{array} \right\rangle \{1,\bar{1},\bar{4}\} \cup \{\bar{2},3\} = \bar{v}_1.$$

The above cycle is displayed in Fig. 2 by double-lined edges. If we would have interchanged $\{2\}$ with $\{3\}$ in $F(v_1 \cap \bar{v}_2)$, and defined v'_2 as above, except $\{2\}$ is replaced by $\{3\}$, then $F(v'_2 \cap \bar{v}_1)$ would be uniquely the the above cycle, except $\{2\}$ and $\{3\}$ would be interchanged. But then we would have $\{1, \bar{1}, \bar{4}\} \cup \{2, \bar{5}\} \in F(v'_2 \cap \bar{v}_1)$ and $\{1, 4, \bar{1}\} \cup \{2, \bar{5}\} \in F(v_1 \cap \bar{v}_2)$ but $\{1, \bar{1}, \bar{2}\} \cup \{2, \bar{5}\} \in F(v_2 \cap \bar{v}_1)$. So axiom (ii) of abstract polytopes would have been violated since $\{1, 2, \bar{1}, \bar{5}\}$ would be a subset of 3 vertices of P.

Hence, we can conclude that $\{2\}$ is not interchangeable with $\{3\}$ in $F(v_1 \cap \bar{v}_2)$ (i.e., the structure of this face which is given above is the only structure which is compatible with $F(v_2 \cap \bar{v}_1)$ given in (a) and the assumption that no path of length less than 4 joined a member of W to a member of $\bar{W}$).

Finally, letting $\bar{v}'_2 = \{1, 3, \bar{1}\} \cup \{\bar{3}, \bar{4}\}$, we consider $F(v_1 \cap \bar{v}'_2) = F(1, 3, \bar{1})$. We note that $\{1, 3, \bar{1}\} \cup \{5, \bar{2}\}$, $\{1, 3, \bar{1}\} \cup \{2, \bar{2}\}$, $\{1, 3, \bar{1}\} \cup \{4, \bar{5}\}$ and $\{1, 3, \bar{1}\} \cup \{\bar{4}, \bar{5}\}$ are already vertices belonging to $F(v_1 \cap \bar{v}_2)$. We get applying (a) that the form of $F(v_1 \cap \bar{v}'_2)$ is necessarily as follows:

$$v_1 = \{1,3,\bar{1}\} \cup \{2,4\} \left\langle \begin{array}{c} \{1,3,\bar{1}\} \cup \{4,\bar{5}\} - \{1,3,\bar{1}\} \cup \{\bar{4},\bar{5}\} - \{1,3,\bar{1}\} \cup \{\bar{3},\bar{4}\} \\ \\ \{1,3,\bar{1}\} \cup \{2,\bar{2}\} - \{1,3,\bar{1}\} \cup \{5,\bar{2}\} - \{1,3,\bar{1}\} \cup \{5,\bar{3}\} \end{array} \right) = \bar{v}_2'.$$

Collecting all the vertices of the four 2-dimensional faces considered above it is not difficult to verify that they form a 3-dimensional abstract polytope which has the structure described in Fig. 2.

Remark 6.3. Klee and Walkup [4] named the property that every path from a member of W to a member of $\bar{W}$ has a length of at most 3, as "Property A". They showed that every 3-dimensional simple polytope with 8 facets satisfies Property A. We have shown that all 3-dimensional abstract polytopes with 8 facets also satisfy Property A except one, namely the one with structure given in Fig. 2. Note, however, that this structure is non-planar. For simple polytopes this cannot happen by Steinitz's theorem [2]. Therefore, noting Theorems 5.3 and 6.1(ii), the above constitutes a new proof Klee and Walkup's theorem for $\Delta_b(2d, d)$ based on simpler assumptions, i.e., that the Hirsch d-step conjecture is true for $d \leq 5$.

Proof of Theorem 6.1 (part (iv) for $d = 5$). It is obvious that (iv) holds if and only if there exists a path of length 5 from v_0 to $\bar{v}_0$, i.e., if and only if there exists a path of length 3 from a neighbor of v_0 to a neighbor of $\bar{v}_0$.

Suppose (iv) does not hold, then by Theorem 6.2(b) every 3-dimensional face of P which is generated by a member of $N(v_0)$ and a member of $N(\bar{v}_0)$ with two symbols in common has the structure of Fig. 2 after relabeling. In particular, let $v_1 = \{1, 2, 3, 4, \bar{1}\}$, $v_2 = \{1, 2, 3, \bar{1}, \bar{2}\}$, $\bar{v}_1 = \{1, \bar{1}, \bar{2}, \bar{3}, \bar{4}\}$ and $P' = F(v_1 \cap \bar{v}_1) = F(1, \bar{1})$ has the form of Fig. 2. Consider $v_0 = \{1, 2, 3, 4, 5\}$ and its incident edge generated by $\{1, 3, 4, 5\}$. The other vertex incident to this edge cannot be $\{1, 3, 4, 5\} \cup \{\bar{i}\}$, where $\bar{i} = \bar{1}, \bar{2}, \bar{3}, \bar{4}$, since this would imply that there is a path of length 5 from v_0 to $\bar{v}_0$ via that edge and one of the following four vertices of P':

$$\{1, 3, 5, \bar{1}, \bar{3}\}, \qquad \{1, 4, 5, \bar{1}, \bar{3}\}, \qquad \{1, 3, 5, \bar{1}, \bar{2}\}, \qquad \{1, 4, 5, \bar{1}, \bar{4}\}.$$

Hence $\{1, 3, 4, 5, \bar{5}\}$ is a vertex adjacent to v_0.

Similar arguments lead to the conclusion that either the set $\{1, 2, 4, 5, \bar{5}\}$ or $\{1, 2, 4, 5, \bar{2}\}$ is the vertex other than v_0 incident to the edge generated by $\{1, 2, 4, 5\}$.

The same argument with respect to $\bar{v}_0$ and the edge generated by $\{\bar{1}, \bar{2}, \bar{4}, \bar{5}\}$ (because of the presence of vertices $\{1, 4, \bar{1}, \bar{2}, \bar{5}\}, \{1, 3, \bar{1}, \bar{4}, \bar{5}\}$, $\{1, 2, \bar{1}, \bar{2}, \bar{5}\}$, $\{1, 2, \bar{1}, \bar{4}, \bar{5}\}$ in P') implies that $\{5, \bar{1}, \bar{2}, \bar{4}, \bar{5}\} \in N_P(\bar{v}_0)$. Consider now vertices $\{1, 3, 4, 5, \bar{5}\}$ and $\{5, \bar{1}, \bar{2}, \bar{4}, \bar{5}\}$, by Theorem 6.2, in order not to have a path of length 3 joining these two vertices, the face of their intersection $F(5, \bar{5})$ must have the structure of Fig. 2. But note in Fig. 2 that the three neighbors of v_1 which are in $N_{P'}^2(v_0)$ have the property that no two are neighbors. This rules out the possibility that $\{1, 2, 4, 5, \bar{5}\}$ is a vertex as it would lie in this face and would be a neighbor of $\{1, 3, 4, 5, \bar{5}\}$ in $N(v_0)$ instead of $N^2(v_0)$. Hence if (iv) does not hold, $\{1, 2, 4, 5, \bar{2}\} \in N(v_0)$.

Let us now consider the face $F(\{1, 2, 4, 5, \bar{2}\} \cap \bar{v}_1\}) = F(1, \bar{2})$. By Theorem 6.2, under the assumption that (iv) does not hold, $F(1, \bar{2})$ must have the structure of Fig. 2. It contains the nonempty 2-dimensional face $F(v_2 \cap \bar{v}_1) = F(1, \bar{1}, \bar{2})$. But note that $F(1, \bar{1}, \bar{2})$ also lies in $F(v_1 \cap \bar{v}_1)$ and has the 7 vertices shown connected by heavy arcs in Fig. 2. But all 2-dimensional faces with seven vertices of the abstract polytope given in Fig. 2 have the property that one of its vertices is adjacent to v_1 in P' and thus analogously one of the seven vertices of $F(1, \bar{1}, \bar{2})$ should be adjacent to $\{1, 2, 4, 5, \bar{2}\}$ in $F\{1, \bar{2}\}$ but in fact none are, a contradiction. So (iv) must hold for $d = 5$.

The last part of Theorem 6.1(iv), for $d \geq 6$, will be proved via Theorems 6.4, 6.5 and 6.7.

Theorem 6.4. *Given* $P \in \mathscr{P}(2d, d)$ *(where* $d \geq 4$*) and paths* (v_0, v_1), $\bar{v}_0$ *satisfying* $|v_i \cap \bar{v}_j| = i + j$ *and* $\bar{v}_2', \bar{v}_2'' \in N^2(\bar{v}_0)$ *satisfying* $|v_1 \cap \bar{v}_2' \cap \bar{v}_2''| = 3$, *then there exists a vertex* $v_2 \in N^2(v_0)$ *such that* $|v_2 \cap \bar{v}_2| = 4$, *where either* $\bar{v}_2 = \bar{v}_2'$ *or* $\bar{v}_2 = \bar{v}_2''$.

Proof. By relabeling for $d \geq 4$, we are assuming

$$v_0 = \{1, \ldots, d\}, \qquad \bar{v}_0 = \{\bar{1}, \ldots, \bar{d}\}, \qquad v_1 = \{1, \ldots, d-1, \bar{1}\},$$

$$\bar{v}_2' = \{1, 2, \bar{1}, \ldots, \bar{d}\} \setminus \{\bar{i}, \bar{j}\},$$

$$\bar{v}_2'' = \{1, 2, \bar{1}, \ldots, \bar{d}\} \setminus \{\bar{k}, \bar{l}\} \qquad (\bar{2} \leq \bar{i}, \bar{j}, \bar{k}, \bar{l} \leq \bar{d}),$$

where $\bar{i}, \bar{j}, \bar{k}, \bar{l}$ are all distinct or $\bar{i} = \bar{k}$ and $\bar{i}, \bar{j}, \bar{l}$ are distinct. Note that $v_1 \cap \bar{v}_2' \cap \bar{v}_2'' = \{1, 2, \bar{1}\}$.

By Theorem 5.1, there exists an $\bar{i}$-avoiding path from v_1 to $\bar{v}_2'$ in $P' = F(v_1 \cap \bar{v}_2')$. Let $Z = N(v_0) \cap P'$. Since $v_1 \in Z$, while $\bar{v}_2' \notin Z$, this path intersects $N_{P'}(Z)$, say at v_2. Note that $v_2 \in N_{P'}(Z) \subset N_P^2(v_0)$.

By definition, all the vertices of $P' = F(v_1 \cap \bar{v}_2')$ contain the symbols $\{1, 2, \bar{1}\}$, and v_2 contains also some $\{\bar{s}\} \in \{\bar{t} : \bar{t} \in \{\bar{2}, \ldots, \bar{d}\}, \bar{t} \neq \bar{i}\}$. But either $\bar{v}_2'$ or $\bar{v}_2''$ must contain $\{\bar{s}\}$. Hence either $\bar{v}_2 = \bar{v}_2'$ or $\bar{v}_2 = \bar{v}_2''$ satisfies $|v_2 \cap \bar{v}_2| = 4$.

Theorem 6.5. *Let* $P \in \mathscr{P}(2d, d)$ *and let* $\{v_0, v_1, v_2\}$, $\{\bar{v}_0, \bar{v}_1\}$ *be two paths in* P *such that* $|v_i \cap \bar{v}_j| = i + j$. *Let* $W = N(v_0) \cap F(v_1 \cap \bar{v}_1)$; *then if* $d \geq 6$ *and* $|W| \geq 2$, *there exists* $v_2' \in N^2(v_0)$ *and* $\bar{v}_2' \in N^2(\bar{v}_0)$ *such that* $|v_2' \cap \bar{v}_2'| = 4$.

Proof. By relabeling, let

$$v_0 = \{1, \ldots, d\}, \qquad v_1 = \{1, \ldots, d-1, \bar{1}\}, \qquad v_2 = \{1, \ldots, d-2, \bar{1}, \bar{2}\};$$

$$\bar{v}_0 = \{\bar{1}, \ldots, \bar{d}\}, \qquad \bar{v}_1 = \{1, \bar{1}, \ldots, \overline{d-1}\}.$$

Define

$$P' = F(v_1 \cap \bar{v}_1), \qquad W = N(v_0) \cap P', \qquad P'' = F(v_2 \cap \bar{v}_1),$$

$$\bar{Z} = N(\bar{v}_0) \cap P'', \qquad \bar{U}_i = \{v \in N_{P''}(\bar{Z}) : \{i\} \subset v\} \quad (i = 2, \ldots, d).$$

Theorem 6.5 is an immediate consequence of the following lemma, because the denial of the existence of such $v_2', \bar{v}_2'$ implies by (b4) below that $|W| = 1$ whereas by hypothesis $|W| \geq 2$.

Lemma 6.6. (a) $\bar{U}_2, \ldots, \bar{U}_d$ *partitions* $N_{P''}(\bar{Z})$. *At least one* $\bar{U}_i \neq \emptyset$, $2 \leq i \leq d - 1$.

(b) *If there exist no* $v_2' \in N^2(v_0)$ *and* $\bar{v}_2' \in N^2(\bar{v}_0)$ *such that* $|v_2' \cap \bar{v}_2'| = 4$, *then*

(b1) $|\bar{U}_i| = 0$ *for* $i = 2, \ldots, d - 2$,
(b2) $|\bar{U}_{d-1}| = 1$,
(b3) $|\bar{U}_d| \geq d - 4$,
(b4) $|W| = 1$ *for* $d \geq 6$.

Proof. (a) Every vertex of $P'' = F(v_2 \cap \bar{v}_1) = F(1, \bar{1}, \bar{2})$ contains $\{1\}$ and thus every vertex of $\bar{Z} = N(\bar{v}_0) \cap P''$ contains exactly one unbarred symbol—namely $\{1\}$. Every vertex of $N_{P''}(\bar{Z})$ contains $\{1\}$ and exactly one other non-barred symbol. Obviously,

$$N_{P''}(\bar{Z}) = \bigcup_{i=2}^{d} \bar{U}_i \quad \text{and} \quad \bar{U}_i \cap \bar{U}_j = \emptyset \quad \text{for } i,j = (2, \dots, d), \ i \neq j.$$

By Theorem 5.1, there exists a $\{d\}$-avoiding path joining v_2 to $\bar{v}_1$ in P''. Since $\bar{v}_1 \in \bar{Z}$ and $v_2 \notin \bar{Z}$, this path must intersect $N_{P''}(\bar{Z})$. Thus, there exists a vertex $\bar{v}_2' \in N_{P''}(\bar{Z}) \subset N^2(\bar{v}_0)$ which does not contain $\{d\}$ implying at least one $|\bar{U}_i| \neq 0$ for $i = 2, \dots, d-1$.

(b1) Assume $\bar{U}_{i_0} \neq \emptyset$ and $\bar{v}_2' \in \bar{U}_{i_0}$ for some i_0, $2 \leq i_0 \leq d-2$, then $\{1, i_0, \bar{1}, \bar{2}\} \subset \bar{v}_2'$. Hence $|v_2 \cap \bar{v}_2'| = 4$. Moreover, $\bar{v}_2' \in N_{P''}(\bar{Z}) \subset N^2(\bar{v}_0)$. This contradicts the hypothesis of (b) and we conclude $|\bar{U}_i| = 0$ for $2 \leq i \leq d-2$.

(b2) From (b1) and the discussion under (a) we conclude $|\bar{U}_{d-1}| \geq 1$. Assume now that $|\bar{U}_{d-1}| \geq 2$ and let $\bar{v}_2', \bar{v}_2'' \in \bar{U}_{d-1}$. Since $\bar{v}_2'$ and $\bar{v}_2''$ both contain $\{1, d-1, \bar{1}\}$, we have $|v_1 \cap \bar{v}_2' \cap \bar{v}_2''| = 3$. Furthermore, $\bar{v}_2', \bar{v}_2'' \in N_{P''}(\bar{Z}) \subset N^2(\bar{v}_0)$, so by Theorem 6.4 there exists $v_2' \in N^2(v_0)$ such that either $|v_2' \cap \bar{v}_2'| = 4$ or $|v_2' \cap \bar{v}_2''| = 4$, contrary to hypothesis of (b). Thus we conclude that $|\bar{U}_{d-1}| = 1$.

(b3) Suppose $|\bar{Z}| = k$. Note that $k \geq 1$ because $\bar{v}_1 \in \bar{Z}$. The vertices of $\bar{Z}$ have the form, $\{1\} \cup \bar{v}_0 \setminus \{\bar{i}\}$ for $\bar{i} \in R$, where R is a subset of k indices of $\{\bar{3}, \dots, \bar{d}\}$.

By the second axiom of abstract polytopes, the subset $\{1\} \cup v_0 \setminus \{\bar{i}, \bar{j}\}$, $(\bar{i} \in R, \bar{j} \notin R, \bar{j} \in \{\bar{3}, \dots, \bar{d}\})$ is contained by two vertices of P''. Thus every vertex of $\bar{Z}$ gives rise to $d - 2 - k$ distinct vertices in $N_{P''}(\bar{Z})$. Therefore, $|N_{P''}(\bar{Z})| = k(d - 2 - k)$. Hence by (a), (b1) and (b2),

$$|N_{P''}(\bar{Z})| = |\bar{U}_d| + 1 = k(d - 2 - k).$$

The last expression implies that

$$0 < k < d - 2 \quad \text{and} \quad |\bar{U}_d| \geq d - 4.$$

(b4) Finally, let us assume that $d \geq 6$ and $|W| \geq 2$, and let $v_1' \in W$ be distinct from v_1. Note that $v \in W$ is of the form $v_0 \cup \{\bar{1}\} \setminus \{i\}$, $i \neq 1$ and that v_1 is obtained by setting $i = d$ and that v_1' is formed by setting $i = i_0$ for some $i_0 \neq 1$ or d. Thus $\{1, d, \bar{1}\} \subset v_1'$. By (b3), either there exists $v_2' \in N^2(v_0)$ and $\bar{v}_2' \in N^2(\bar{v}_0)$ such that $|v_2' \cap \bar{v}_2'| = 4$ or $|\bar{U}_d| \geq d - 4 \geq 2$

for $d \geq 6$. Accordingly, let $\bar{v}'_2, \bar{v}''_2 \in \bar{U}_d$. Since both $\bar{v}'_2$ and $\bar{v}''_2$ contain $\{1, d, \bar{1}\}$, we have $|v'_1 \cap \bar{v}'_2 \cap \bar{v}''_2| = 3$ which by Theorem 6.4 implies that there exists a $v'_2 \in N^2(v_0)$ such that either $|v'_2 \cap \bar{v}'_2| = 4$ or $|v'_2 \cap \bar{v}''_2| = 4$, contrary to hypotheses of part (b). We conclude for $d \geq 6$ that $|W| = 1$.

Theorem 6.7. *If $d \geq 6$ and if there exist $v'_1, v''_1 \subset N(v_0)$ and $\bar{v}_1 \subset N(\bar{v}_0)$ such that $|v'_1 \cap v''_1 \cap \bar{v}_1| = 2$, then Theorem 6.1(iv) holds.*

Proof. Without loss of generality we can assume that

$$v_0 = \{1, \ldots, d\}, \qquad v_1 = \{1, \ldots, d-1, \bar{1}\},$$
$$v'_1 = \{1, \ldots, d-2, d, \bar{1}\},$$
$$\bar{v}_0 = \{\bar{1}, \ldots, \bar{d}\}, \qquad \bar{v}_1 = \{1, \bar{1}, \ldots, \overline{d-1}\},$$
$$P' = F(v_1 \cap \bar{v}_1), \qquad W = P' \cap N(v_0).$$

We wish to show that if $d \geq 6$, then there exist $v'_2 \in N^2_P(v_0)$ and $\bar{v}'_2 \in N^2_P(\bar{v}_0)$ such that $|v'_2 \cap \bar{v}'_2| = 4$.

By Theorem 5.1, there exists a $\{\bar{d}\}$-avoiding path from v_1 to $\bar{v}_1$ in $P' = F(v_1 \cap \bar{v}_1) = F(1, \bar{1})$. This path intersects $N_{P'}(W)$ at v_2 (say). In this case, $v_0, v_1, v_2, \bar{v}_0, \bar{v}_1$ satisfy the conditions of Theorem 6.5. Moreover, $|W| \geq 2$ since $v_1, v'_1 \subset W$ and $d \geq 6$ so that by Theorem 6.5 there exist $v'_2 \in N^2_P(v_0)$ and $\bar{v}'_2 \in N^2_P(\bar{v}_0)$ such that $|v'_2 \cap \bar{v}'_2| = 4$.

Corollary 6.8. *If $d \geq 6$ and if there exist adjacent $v'_1, v''_1 \subset N(v_0)$, then Theorem 6.1(iv) holds.*

Proof. Let $\{\bar{1}\} \subset v'_1 \cap v''_1$ and let $\{k\} \notin v'_1$ and $\{\ell\} \notin v''_1$. For $i = (2, \ldots, d)$, let $\bar{u}_i \in N(\bar{v}_0)$ such that vertex $\bar{u}_i \supset \bar{v}_0 \setminus \{\bar{i}\}$. Let $p_i = \bar{u}_i \cap v_0$. If for some $i, p_i \notin \{k, \ell\}$, then $\{\bar{u}_i \cap v'_1 \cap v''_1\} = \{\bar{1}, p_i\}$ and Theorem 6.7 applies. Otherwise, all $\bar{u}_i$ contain $\{k\}$ or $\{\ell\}$, and there exist a $\bar{u}_s$ and $\bar{u}_t$, both of which contain $\{k\}$ (say), but then $\{\bar{u}_s \cap \bar{u}_t \cap v''_1\} = \{\bar{1}, k\}$ and again Theorem 6.7 applies.

Proof of Theorem 6.1 (part (iv) for $d \geq 6$). By Theorem 6.1(iii), we can assume the existence of paths $(v_0, v_1, v_2), (\bar{v}_0, \bar{v}_1)$ such that $|v_i \cap \bar{v}_j| = i + j$. Without loss of generality we can assume that

$$v_0 = \{1, \ldots, d\}, \quad v_1 = \{1, \ldots, d-1, \bar{1}\}, \quad v_2 = \{1, \ldots, d-2, \bar{1}, \bar{2}\},$$

$$\bar{v}_0 = \{\bar{1}, \ldots, \bar{d}\}, \quad \{\bar{v}_1 = 1, \bar{1}, \ldots, \overline{d-1}\}.$$

Let us define P', P'', W, $\bar{Z}$ and $\bar{U}_i$, $(i = 2, \ldots, d)$ as in Lemma 6.6. Since we assume that $d \geq 6$, we have by Lemma 6.6 that $|\bar{U}_d| \geq 2$. Let $\bar{v}_2, \bar{v}_2' \in \bar{U}_d$.

If $|\bar{Z}| \geq 2$, then (considering the two vertices in $\bar{Z}$ and v_1) (iv) holds by Theorem 6.7. If $|\bar{Z}| = 1$, then $\bar{Z} = \{\bar{v}_1\}$ and necessarily $\bar{v}_2, \bar{v}_2'$ have the form

$$\bar{v}_2 = \{1, d, \bar{1}, \ldots, \overline{d-1}\} \setminus \{\bar{i}_0\},$$

$$\bar{v}_2' = \{1, d, \bar{1}, \ldots, \overline{d-1}\} \setminus \{\bar{j}_0\}$$

for some $i_0, j_0, \bar{3} \leq \bar{i}_0, \bar{j}_0 \leq \overline{d-1}$ and $\bar{i}_0 \neq \bar{j}_0$.

Let $W' = F(1) \cap N(v_0)$. Note $|W'| = d - 1$. Every $v \in W'$ contains $1, d$, except v_1. If any $v_1' \in \{W' \setminus v_1\}$ contains $\bar{i} \notin \{\bar{i}_0, \bar{j}_0, \bar{d}\}$, then $|v_1' \cap \bar{v}_2 \cap \bar{v}_2'| = 3$ so that Theorem 6.1(iv) follows from Theorem 6.4. If, on the contrary, all $v \in \{W' \setminus v_1\}$ contain either $\bar{d}$ or $\bar{i}_0$ or $\bar{j}_0$, then there exists a pair $v_1', v_1'' \in \{W' \setminus v_1\}$ both of which contain $\{1, \bar{d}\}$ or $\{1, \bar{i}_0\}$ or $\{1, \bar{j}_0\}$ because $|W'v_1| = d - 2 \geq 4$ for $d \geq 6$. We may now apply Corollary 6.8.

Theorem 6.9. (i) $\Delta_a(2d + 1, d) \leq \Delta_a(2d, d - 1) + 1$, *for* $d \geq 2$,

(ii) $\Delta_a(2d, d) \leq \Delta_a(2d - k, d - k) + k$, *for* $k = (1, 2, 3, 4), d - k \geq 2$.

Proof. (i) Let $P \in \mathscr{P}(2d + 1, d)$ such that $\delta P = \Delta_a(2d + 1, d)$ and let the minimum path joining v_0 to $\bar{v}_0$ in P has length $\Delta_a(2d + 1, d)$. By Theorem 5.3, we can assume $v_0 \cap \bar{v}_0 = \emptyset$ and there exists $v_1 \in N(v_0)$ such that $|v_1 \cap \bar{v}_0| = 1$, otherwise all $v \in N(v_0)$ would be neighbors and there would be no path from v_0 to $\bar{v}_0$. The result follows since $\delta[F(v_1 \cap \bar{v}_0)] \leq \Delta_a(2d, d - 1)$.

(ii) Follows immediately from Theorem 6.1.

Remark 6.10. *Relations for simple polytopes.* Note that the various arguments presented apply if the phrase "simple polytope" is substituted for abstract polytope wherever it occurs, and the term $\Delta_a(n, d)$ is replaced by $\Delta_b(n, d)$ (the maximum diameter of ordinary polytopes over all d-dimensional polytopes with n facets) and therefore the various theorems and corollaries are also valid after the replacement of these terms.

7. Maximum diameters of abstract polytopes and the Hirsch conjecture

Corresponding to the Hirsch conjecture of simple polytopes is the conjecture for abstract polytopes that

$$\Delta_a(n, d) \le n - d \qquad (d > 1, n \ge d + 1).$$

Theorem 7.1 is the analog of the results of Klee and Walkup [4] for abstract polytopes (except for $\Delta_b(n, 3) = [2n/3] - 1$ for $n \ge 9$) and is mainly based on Theorem 6.1.

Theorem 7.1. *The values of $\Delta_a(n, d)$ for $n - d \le 5$, and all d are as given in Table 1. In addition, $\Delta_a(n, 2) = [n/2]$.*

Table 1
Values of $\Delta_a(n,d)$

$n-d$ d	1	2	3	4	5	
1	1	×	×	×	×	
2	1	2	2	3	3	$\ldots \Delta_a(n,2) = [n/2]$
3	1	2	3	3	4	
4	1	2	3	4	5	
≥ 5	1	2	3	4	5	

Proof. Let $P \in \mathscr{P}(n, d)$, $\delta P = \Delta_a(n, d)$. By Theorem 5.3, we can further assume for $n \ge 2d$, that there exist $v_0, \bar{v}_0 \in P$ such that $v_0 \cap \bar{v}_0 = \emptyset$ and $\rho(v_0, \bar{v}_0) = \Delta_a(n, d)$.

(a) $2d > n$. By Theorem 5.2(iv), each column of Table 1 is constant from the main diagonal downwards.

(b) $d = 2, n \ge 4$. Since P is a 2-dimensional abstract polytope, the number of vertices of P is equal to the number of its edges, therefore the graph of P forms a simple cycle with n vertices. Hence $\Delta_a(n, d) = [n/2]$.

(c) $n = 2d, d \le 5$. Applying Theorem 6.1, $\Delta_a(2d, d) = \rho(v_0, \bar{v}_0) = d$.

(d) $d = 3, n = 7$. Let $\bigcup P \setminus \{v_0 \cup \bar{v}_0\} = A$; then by Theorem 5.1 there exists an A-avoiding path between v_0 and $\bar{v}_0$. This path intersects $N^2(v_0)$ at v_2 (say). Since every vertex in $N^2(v_0)$ contains two symbols of $\{\bigcup P \setminus v_0\}$, v_2 is necessarily adjacent to $\bar{v}_0$. Hence $\Delta_a(7, 3) \le 3$. Since $\Delta_a(7, 3) \ge \Delta_a(6, 2) = 3$, by Theorem 5.2, we obtain $\Delta_a(7, 3) = 3$.

(e) $d = 3$, $n = 8$. Let $\bigcup P = \{1, 2, 3, 4, 5, 6, 7, 8\}$, $v_0 = \{1, 2, 3\}$, $\bar{v}_0 = \{4, 5, 6\}$, where $\Delta_a(8, 3) = \rho(v_0, \bar{v}_0)$. Fig. 2 is an abstract polytope belonging to $\mathscr{P}(8, 3)$ with diameter $\delta = 4$. Therefore, $\delta(P) \geq 4$. Assume $\delta(P) > 4$, then every vertex contains either $\{7\}$ or $\{8\}$ for otherwise a vertex in $N(v_0)$ and $\bar{v}_0$ (or in $N(\bar{v}_0)$ and v_0) would both contain a symbol in common, say $\{5\}$, and we would have $\Delta_a(8, 3) = \delta(P) \leq 1 + \delta(F(5)) \leq 1 + \Delta_a(7, 2) = 4$. Thus we can assume without loss of generality $N(v_0) = \{1, 2, 7\}; \{1, 3, 7\}; \{2, 3, 8\}$ and $N(\bar{v}_0) = \{4, 5, 7\}; \{4, 6, 8\};$ and either $\{5, 6, 7\}$ or $\{5, 6, 8\}$. Consider now the cycle $F(7)$ which can contain at most seven vertices. In the first case, the shorter leg of the cycle joining $N(v_0)$ to $N(\bar{v}_0)$ provides a path of length 2. In the second case, neither $\{4, 6, 7\}$ nor $\{5, 6, 7\}$ can appear in the cycle so that it has at most six vertices and it too provides a path of length 2. Thus $4 \leq \Delta_a(8, 3) = \rho(v_0, \bar{v}_0) \leq 4$.

(f) $d = 4$, $n = 9$. Klee and Walkup [4] exhibit a $P \in \mathscr{P}(9, 4)$ with $\delta P = 5$. Thus, by Theorem 6.9 and (e), $5 \leq \Delta_a(9, 4) \leq \Delta_a(8, 3) + 1 = 5$.

References

[1] I. Adler, G.B. Dantzig and K. Murty, Existence of A-avoiding paths in abstract polytopes, *Mathematical Programming Study* 1 (1974) 41–42.
[2] B. Brunbaum, *Convex polytopes* (Wiley, New York, 1967).
[3] G.B. Dantzig, *Linear programming and extensions* (Princeton University Press, Princeton, N.J., 1963).
[4] V. Klee and D.W. Walkup, The d-step conjecture for polyhedra of dimension $d < 6$, *Acta Mathematica* 117 (1967) 53–78.

Mathematical Programming Study 1 (1974) 41–42. North-Holland Publishing Company

EXISTENCE OF A-AVOIDING PATHS
IN ABSTRACT POLYTOPES*

Ilan ADLER** and George DANTZIG

Stanford University, Stanford, Calif., U.S.A.

and

Katta MURTY

University of Michigan, Ann Arbor, Mich., U.S.A.

Received 29 November 1971

Theorem. *Let P be an abstract polytope and let $A \in \bigcup P$. If v, $\bar{v} \in P \setminus F_P(A)$ (i.e., if both v and $\bar{v}$ do not contain A), then there exists a path (called a A-avoiding path) joining v and $\bar{v}$ in $G(P)$ (the graph of P) such that no vertex of that path belongs to $F_P(A)$.*

Abstract polytopes include as special cases simple convex polytopes. The latter can be represented as non-degenerate bounded feasible linear programs. The analoguous theorem for convex polytopes states: If two feasible bases have a column in common, then it is possible to pass from one to the other via a sequence of adjacent feasible basis changes all of which have the column in common. The usual proof is to assign as linear objective the maximization of the variable of the common column. Then two paths exist one from each of the initial feasible bases to an optimal basis. All bases along each path have the column in common. Hence if there is a unique optimal basis, the paths may be joined together at the optimum point to form a simple connecting path.

---

* All definitions and notation used here are given in [1] (the preceding paper).
** Presently, University of California, Berkeley, Calif., U.S.A.

It is not hard to extend the result to the non-unique case as well. For abstract polytopes, however, there is no objective function, hence a different type of proof is required.

Proof. Let P be a d-dimensional abstract polytope.

(a) $d \leq 1$. The proof is trivial.

(b) $d = 2$. By axiom (ii), $G(P)$ forms a simple cycle whose edges correspond to the facets of P. Obviously, removing the edge A from $G(P)$ cannot disconnect $G(P)$.

(c) $d \geq 3$. Let $P' = F_P(A)$ and let $v, \bar{v} \in P \setminus P'$. By axiom (iii), there exists a sequence of adjacent vertices $v = v_0, \ldots, v_k = \bar{v}$. By axiom (ii), if $v_i \in P'$, then there exists a unique vertex $\bar{v}_i$ such that $\bar{v}_i$ is a neighbor of v_i and $\bar{v}_i \notin P'$. Let

$$u_i = \begin{cases} v_i & \text{if } v_i \notin P' \\ \bar{v}_i & \text{if } v_i \in P' \end{cases} \quad (i = 1, \ldots, k).$$

(Note that the u_i need not be distinct.) Since $|u_i \cap u_{i+1}| \geq d - 2$ $(i = 1, \ldots, k)$, $\{u_i \cap u_{i+1}\}$ generates a face of P of dimension 0 (a vertex), or 1 (an edge), or 2 (a cycle), thus there exists, by (a)–(b), an A-avoiding path p_i on $F_P(\{u_i \cap u_{i+1}\})$ joining u_i and u_{i+1} $(i = 1, \ldots, k)$. Hence $\bigcup_{i=1}^{k} p_i$ is an A-avoiding path in P joining v to $\bar{v}$.

Reference

[1] I. Adler and G.B. Dantzig, "Maximum diameter of abstract polytopes", *Mathematical Programming Study* 1 (1974) 20–40.

Mathematical Programming Study 1 (1974) 43–58. North-Holland Publishing Company

ON TWO SPECIAL CLASSES
OF TRANSPORTATION POLYTOPES*

M.L. BALINSKI

Graduate Division and University Center
City University of New York, New York, N.Y., U.S.A.
and
Ecole Polytechnique Federale de Lausanne, Lausanne, Switzerland

Received 8 November 1973
Revised manuscript received 6 April 1974

Dedicated to A.W. Tucker with affection, gratitude, and respect

The Hirsch conjecture on the diameter of polytopes, and the form and number of vertices are here established constructively for two special classes of transportation polytopes.

0. Introduction

A transportation polytope is defined by

$$P_{m,n}(\{a_i\}; \{b_j\}) = P_{m,n}(a_1, \ldots, a_m; b_1, \ldots, b_n)$$

$$= \{x = (x_{ij}): \textstyle\sum_j x_{ij} = a_i, \sum_i x_{ij} = b_j, x_{ij} \geq 0\},$$

where $a_i > 0$, $b_j > 0$ for all i and j and $\sum_1^m a_i = \sum_1^n b_j$.

Klee and Witzgall [8] studied such polytopes, concerning themselves primarily with characterizing and counting the facets and vertices. They left virgin—or at least uncertain—the status of the Hirsch conjecture [5, 7] which states that to go from any one vertex of $P_{m,n}$ to any other, all the while traveling on the extreme edges, at most $m + n - 1$ such edges need be traversed (see Section 3 for a summary discussion of the

* This work was supported by the Army Research Office, Durham, under contract No. DA-31-124-ARO(D)-366.

Hirsch conjecture). They found, for example, that any $P_{m,n}$ has at least $n!/(n - m + 1)!$ extreme points with $P_{m,n}(1, \ldots, 1, n - m + 1; \{1\})$ having precisely that number. They conjectured that $P_{m,n}(\{n\}, \{m\})$, for $(m, n) = 1$, contains the maximum number of extreme points or vertices of any polytope $P_{m,n}$, and computed the numbers of vertices for the two special cases $n = \mu m + 1$ and $n = \mu n - 1$. This conjecture has since been established by Bolker [4].

This paper considers two special classes of transportation polytopes which are generalizations of the two special cases of Klee and Witzgall, counts the number of vertices of each (in one case via an argument considerably simpler than that found in [8]), and constructively establishes the Hirsch conjecture for these two classes.

In what follows, $C(I, J)$ is the complete bi-partite graph with node-set $I \cup J, |I| = m, |J| = n$, and edge-set $(i, j), i \in I, j \in J$. Given any $x \in P_{m,n}(\{a_i\}; \{b_j\})$, associate the sub-graph $G(x)$ of $C(I, J)$ defined by: (i, j) is an edge of $G(x)$ if and only if $x_{ij} \neq 0$. Then $x \in P_{m,n}$ is a vertex or extreme point if and only if $G(x)$ contains no cycle. A graph containing no cycle is a *forest*; a connected forest is a *tree*. Thus $G(x)$, for x a vertex of $P_{m,n}$, is a forest, cannot contain more than $m + n - 1$ edges, and is a (spanning) tree if it contains precisely $m + n - 1$ edges. Distinct vertices of P have distinct graphs, so "spanning" forests (meaning each node has valency at least 1) which possess additional properties provide a perfect model for vertices of $P_{m,n}$. If, for x a vertex of $P_{m,n}$, $G(x)$ is a forest and not a tree, then such an x is called "degenerate". It "corresponds" to a situation in which a partial sum of the a_i is equal to a partial sum of the b_j. Two vertices $x, y \in P_{m,n}$ are *neighbors*, or are connected by a 1-dimensional face of $P_{m,n}$, if and only if the union of the edge sets of $G(x)$ and $G(y)$ contains exactly one cycle. The forests $G(x)$ and $G(y)$ are then also said to be neighbors. Thus one can speak of "going from x to y" along an extreme edge by adjoining one edge to and dropping one edge from the cycle thereby formed in $G(x)$, to obtain $G(y)$, and this can be referred to as constituting "one-step". These facts are all easily established by elementary argument (see, for example, [3, 5, 8]).

Going from one vertex $x \in P_{m,n}$ to a neighboring vertex y corresponds precisely to making a *pivot* step from x to y, and in this sense this paper uses the pivoting idea to establish a geometric property. This attests, once again, to A.W. Tucker's deep insight in insisting upon the fundamental nature of the pivotal operation as a constructive computational *and* mathematical device. At the same time it permits me to express my personal indebtedness and warm thanks for his initial guidance, when he encouraged me in this line of thinking, for his continuing interest and counsel in professional endeavors, in a word, for his role as teacher and friend to me.

1. $P_{m,n}(\{k_i m + 1\}; \{m\})$

Consider the class of polytopes $P_{m,n}(\{k_i m + 1\}; \{m\})$, where $k_i \geqslant 0$ and integer. Then n is defined by $mn = \sum_1^m (k_i m + 1)$, or, taking $\bar{k} = \sum_1^m k_i, n = \bar{k} + 1$. Note that there exist such polytopes corresponding to any chosen values of m and n. Since no partial sum of $a_i = k_i m + 1$ is equal to a partial sum of $b_j = m$, a vertex x of $P_{m,n}(\{k_i m + 1\}; \{m\})$ has as its graph $G(x)$ which must be a spanning tree of $C(I, J), |I| = m$, $|J| = \bar{k} + 1$, and hence contains precisely $m + \bar{k}$ edges.

Define the particular family of spanning trees of $C(I, J)$, $\mathcal{T}_m^v(\{k_i\})$, called *trees of constant source valency*, to be trees having valency exactly $k_i + 1$ at each node $i \in I$.

Theorem 1.1. *The trees $\mathcal{T}_m^v(\{k_i\})$ are in one-one correspondence with the extreme points of $P_{m,n}(\{k_i m + 1\}; \{m\})$, where $n = \bar{k} + 1 = \sum_i k_i + 1$.*

Proof. First, consider an arbitrary vertex x of P. (We drop the various identifying parameters wherever no confusion can arise). Since no x_{ij} can exceed m, every row i of $x = (x_{ij})$ must have at least $k_i + 1$ positive x_{ij}. Therefore, the $G(x)$-valency of i is at least $k_i + 1$, whence the number of edges of $G(x)$ at least $\sum_i (k_i + 1) = \bar{k} + m$. But $G(x)$ contains precisely $\bar{k} + m$ edges, so each $G(x)$-valency of $i \in I$ is precisely $k_i + 1$.

Second, consider any $T \in \mathcal{T}_m^v$ and any edge (i, j) of T. The removal of (i, j) from T leaves two components, one of which T_{ij} contains $i \in I$. Define x_{ij} to be the number of sources of T_{ij}. Then

$$\sum_j x_{ij} = k_i m + 1$$

because source i is counted $k_i + 1$ times, whereas each other source—which is excluded from the sum exactly once, there being a unique path from i to any node in T—is counted exactly k_i times, giving $k_i + 1 + (m - 1) k_i = k_i m + 1$. On the other hand,

$$\sum_i x_{ij} = m$$

since each source is counted exactly once in this sum. Thus, $T \in \mathcal{T}^v$ admits $x \in P$ with $x_{ij} > 0$ only if (i, j) is an edge of T, and x must be an extreme point of P with $T = G(x)$.

A *claw* C_i on $i \in I$ in $C(I, J)$ is a tree having one node $i \in I$ and any node $j \in J$ with valency at most 1. The cardinality of a claw, $|C_i|$, is its number of edges. A set of claws are *independent* if they have no nodes in common. A *rooted set of claws* $(r, \{C_i\})$ with root $r \in J$ is a collection of independent claws, one C_i on each $i \in I$, with $|C_i| = k_i$ such that each $j \in J \sim \{r\}$ belongs to exactly one C_i. Note that $|C_i| = k_i = 0$ is possible, and root r belongs to none of the claws in $(r, \{C_i\})$.

Lemma 1.2. *If $T \in \mathcal{T}^v$ and some arbitrary $r \in J$ is singled out as the root, then T contains a unique rooted set of claws $(r, \{C_i\})$.*

Proof. There exists a unique path joining r to each $i \in I$ in T. Let $(i, j(i, r))$ be the unique edge on this path incident to i. Drop these m edges from T to obtain the forest F containing $\bar{k}$ edges. Clearly, every node $j \in J$, $j \neq r$ is of F-valency 1, and every node $i \in I$ is of F-valency k_i. Thus (r, F) is a rooted set of claws. F is unique, given r, for suppose (r, F') was also a rooted set of claws, $F \neq F'$. Then there exists $j_0 \in J$ with $(i_0, j_0) \in F$ and $\notin F'$. So there exist i_1 and j_1 with $(i_1, j_0) \in F'$ and $\notin F$, $(i_0, j_1) \in F'$ and $\notin F$. Continued application of this observation shows $F \cup F'$ contains a cycle, a contradiction.

Lemma 1.3. *Consider $C(I^*)$, the complete graph on $|I^*| = m + 1$ nodes, $m \geq 2$. $C(I^*)$ contains a spanning tree with node i having valency $\ell_i + 1$ if and only if $\sum_i \ell_i = m - 1$, $\ell_i \geq 0$ and integer. Moreover, there exist precisely*

$$\binom{m - 1}{\ell_1, \ell_2, \dots, \ell_{m+1}}$$

such trees. (The case $m = 1$ is trivial).

Proof. If T spans $C(I^*)$, then it contains m edges, so $\sum_i (\ell_i + 1) = 2m$ or $\sum_i \ell_i = m - 1$.

On the other hand, let $\{\ell_i\}$ be any nonnegative integers summing to $m - 1$, order them (for convenience) $\ell_1 \geq \ell_2 \geq \dots \geq \ell_{m+1}$ and then proceed via induction. At least $\ell_m = \ell_{m+1} = 0$. The existence of a tree with valencies $\{\ell_i + 1\}$ is subsumed by the following argument. Define k by $\ell_k > 0, \ell_{k+1} = 0$. $(i, m + 1)$ can belong to a tree having the requisite valencies if and only if $\ell_i > 0$, i.e., $i \leqslant k$. There are, inductively, exactly

$$\binom{m - 2}{\ell_1, \ldots, \ell_i - 1, \ldots, \ell_k},$$

where $\sum_1^k \ell_i = m - 1$, such trees containing $(i, m + 1)$. Therefore, the total number of trees with valencies $\{\ell_i + 1\}$ is

$$\sum_{i=1}^k \binom{m - 2}{\ell_1, \ldots, \ell_i - 1, \ldots, \ell_k} = \binom{m - 1}{\ell_1, \ldots, \ell_k} = \binom{m - 1}{\ell_1, \ldots, \ell_{m+1}}.$$

Theorem 1.4. $|\mathcal{T}_m^v(\{k_i\})| = \bar{k}! \, (\bar{k} + 1)^{m-1} / \prod_1^m k_i!$, where $\bar{k} = \sum_1^m k_i$.

Proof. The approach is to count the total number of rooted sets of claws; and then the total number of trees $T \in \mathcal{T}^v$ corresponding to each rooted set of claws. By Lemma 1.2, this means that every $T \in \mathcal{T}^v$ will have been counted exactly $n = \bar{k} + 1$ times.

There are $(\bar{k} + 1)! / \prod k_i!$ rooted sets of claws.

Now consider any $T \in \mathcal{T}^v$ containing the rooted set of claws $(r, \{C_i\})$. Define $\bar{T} = T/(\{C_i\})$ on $C(I \cup \{r\})$ by

$$(i, r) \in \bar{T} \quad \text{if} \quad (i, r) \in T;$$
$$(i, j) \in \bar{T} \quad \text{if either } (i, k) \in T \text{ with } k \in C_j,$$
$$\text{or} \quad (j, k) \in T \text{ with } k \in C_i.$$

The "or" is clearly exclusive. $\bar{T}$ is a spanning tree on the $m + 1$ nodes $I \cup \{r\}$, for it contains m edges (all edges "of" T save those $\bar{k}$ which correspond to edges of the claws $\{C_i\}$) and there exists a path in $\bar{T}$ connecting every pair of vertices of $I \cup \{r\}$ (otherwise T would not be spanning). The valencies $\{\ell_i + 1\}$, $i = 1, \ldots, m$, and $\ell_r + 1$ of $\bar{T}$ are determined by the valencies of T:

$$\ell_r + 1 \text{ is the } T\text{-valency of } r \text{ in } C(I, J),$$
$$\ell_i + 1 = 1 + \sum_j \{T\text{-valency of } j \in J; (i, j) \in C_i\} - k_i.$$

On the other hand, given any $\bar{T}$ on $C(I \cup \{r\})$ with given valencies

$(\{\ell_i + 1\})$, the following distinct T satisfy $\bar{T} = T/(\{C_i\})$ for T containing the rooted set of claws $(r, \{C_i\})$:

If $(i, j) \in \bar{T}, j \neq r$, is the edge adjacent to i on the unique $\bar{T}$-path joining i to r, then $(i, k) \in T$ for some $k \in C_j, |C_j| = k_j$;

if $(i, r) \in T$, then $(i, r) \in T$.

Therefore, there are $\prod_1^m k_i^{\ell_i}$ trees $T \in \mathcal{T}^v$ containing $(r, \{C_i\})$ such that $\bar{T} = T/(\{C_i\})$. But there are

$$\binom{m-1}{\ell_1, \ldots, \ell_m, \ell_r}$$

trees $\bar{T}$ with valencies $\{\ell_i + 1\}$. Thus, to any rooted set of claws there correspond

$$\sum_\ell \binom{m-1}{\ell_1, \ldots, \ell_m, \ell_r} \prod_1^m k_i^{\ell_i} \cdot 1^{\ell_r} = (1 + \sum k_i)^{m-1} = (1 + \bar{k})^{m-1}$$

trees $T \in \mathcal{T}^v$. This yields

$$(1 + \bar{k}) |\mathcal{T}^v| = \frac{(1 + \bar{k})^{m-1}(\bar{k} + 1)!}{\prod_k^m k_i!},$$

proving the theorem.

The problem which remains (here) to be addressed is the Hirsch conjecture for arbitrary $P_{m,n}(\{k_i m + 1\}; \{m\})$. The essential tool is the one-one correspondence between extreme points of P and trees of $\mathcal{T}^v(\{k_i\})$. Two trees of $\mathcal{T}^v$ are neighbors if and only if their union contains exactly one cycle. So, given $T \in \mathcal{T}^v$, a neighbor T' is obtained by adjoining an edge $(i, j) \notin T$ and then dropping the unique edge in the cycle formed which is incident to i. It is this structural property of the graph of any extreme point which permits the proof which follows. It should be noted, however, that it is *not* possible (necessarily) to find a path of neighbors from any T to any T^* in $\mathcal{T}^v$ in which "no errors are made", i.e., in which *each* step leads to a neighbor containing more edges in common with T^*. For example, take $\mathcal{T}_2^v(1,1)$ (see Fig. 1). It might be hoped that since $|T \cap T^*| = 2$, at most two steps are required, i.e., one intermediate neighbor is sufficient. In fact, it is easily seen that three are required. So the "greedy" approach does not work. Instead what is shown does work is a two stage procedure: in the first stage, (Lemma 1.6) of "going from T to T^* in $\mathcal{T}^{\vee}$", greediness of one particular kind is rewarded and in the second stage (Lemma 1.7) a second kind of greediness is rewarded (this is a moderately greedy approach).

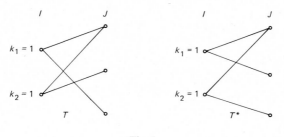

Fig. 1.

Before proceeding greedily, the following, needed in the sequel, is established. An *end-sink* $j \in J$ for a spanning tree T of $C(I,J)$ is a node $j \in J$ of T-valency 1.

Lemma 1.5. *Let δ be the number of end-sinks of any $T \in \mathcal{T}^{\vee}(\{k_i\})$. For $m \geqslant 2$, $\bar{k} - m + 2 \leqslant \delta \leqslant \bar{k}$. Furthermore, if $k_i \geqslant 1$ for all $i \in I$, then at least two source nodes $i \in I$ each have k_i adjacent end-sinks in T.*

Proof. Clearly $\delta \leqslant \bar{k}$ for $m \geqslant 2$. With δ as defined, exactly $\bar{k} + 1 - \delta$ sinks are not end-sinks. Thus, the m source nodes I and the $\bar{k} + 1 - \delta$ sink nodes in question contain a spanning tree having $m + \bar{k} - \delta$ edges of T. But each such sink node has valency at least 2 so that

$$2(\bar{k} + 1 - \delta) \leqslant m + \bar{k} - \delta,$$

establishing the lower bound.

Finally, if, when $k_i \geqslant 1$ all $i \in I$, the last statement did not hold, then at least $m - 1$ source nodes $\bar{I}$, $I = \bar{I} \cup \{i^*\}$, would have $k_i - 1$ adjacent end-sinks or less. Therefore, the greatest number of end-sinks would be

$$\sum_{\bar{I}} (k_i - 1) + k_{i^*} = \bar{k} - m + 1 < \bar{k} - m + 2,$$

a contradiction.

Lemma 1.6. *Given any* $T \in \mathcal{T}^v_m(\{k_i\})$, $k_i \geqslant 1$ *for all* i, *and any set of independent claws* $(\{C^*_i\})$, $|C^*_i| = k_i$, *some* $T^* \in \mathcal{T}^v_m(\{k_i\})$ *containing* $(\{C^*_i\})$ *can be obtained from* T *in* $\sum k_i - \sum |T \cap C^*_i| \leqslant \bar{k}$ *steps.*

Proof. The proof is by induction on m. Let $C_i = T \cap C^*_i$. If it is not true that $C_i = C^*_i$ for all i, then it is shown that it is possible to adjoin an edge of some $C^*_i \sim C_i$ to obtain a neighbor T' of T without dropping any edge of C^*_i. Several cases are distinguished.

If $|C_i| < |C^*_i|$ and $|C_i| = 0$, then any edge of C^*_i may be adjoined to T to obtain a T' containing more edges of $(\{C^*_i\})$.

Assume, therefore, that $|C_i| \geqslant 1$ for all i. At least one C_i, say C_h, has the property that every one of its edges are adjacent to end-sinks in T. For otherwise, every $i \in I$ whose T-valency is at least 2 has at least one adjacent sink of valency at least 2, implying the existence of a cycle.

If $C_h \neq C^*_h$, then adjoining any arc of $C^*_h \sim C_h$ to T obtains a T' which is better since no arc of C_h can belong to the unique cycle formed.

If $C_h = C^*_h$, then consider the tree $\bar{T}$ formed from T by dropping all arcs (h, j), $j \in J$. $\bar{T} \in \mathcal{T}^v_{m-1}(\{k_i\}_{i \neq h})$. By induction, the lemma holds and thus a tree $\bar{T}^*$ containing $(\{C^*_i\}_{i \neq h})$ can be obtained "without mistake". Since no arc of form (h, j) can be affected, were the same steps performed on T, the proof is completed.

Lemma 1.7. *If* $T, T^* \in \mathcal{T}^v_m(\{k_i\})$, $k_i \geqslant 1$ *for all* i *have a common set of independent claws* $(\{C^*_i\})$, $|C^*_i| = k_i$, *then* T^* *can be obtained from* T *in* $|T^*| - |T \cap T^*| \leqslant m$ *steps.*

Proof. The proof again uses induction on m, and again it is shown that it is always possible to adjoin a "wanted" edge of T^* to T, and drop no edge of T^*, to obtain a "better" neighbor T' of T.

Choose any C^*_h containing k_h end-sinks in T. Let (h, t) be the edge of T not belonging to C^*_h, and (h, t^*) be the edge of T^* not belonging to C^*_h. Either $t \neq t^*$ or $t = t^*$.

If $t \neq t^*$, adjoin (h, t^*) to T. The edge $(h, t) \notin T^*$ must be dropped to obtain a neighbor T' since no edge $(h, j) \in C^*_h$ can belong to the cycle formed.

If $t = t^*$, let the common set of $k_h + 1$ nodes $j \in J$ incident to h in both T and T^* be called J_h. Let $\bar{I} = I \sim \{h\}$, $\bar{J} = J \sim J_h \cup \{j_h\}$, and note $|J_h| = k_h + 1$. Consider now $\bar{T}$ and $\bar{T}^*$ obtained, respectively, from T and T^* as follows:

$(i, j) \in T$, respectively T^*; $i \in \bar{I}, j \notin J_h$ implies $(i, j) \in \bar{T}$, respectively $\bar{T}^*$, and

$(i, j) \in T$, respectively T^*; $i \in \bar{I}, j \in J_h$ implies $(i, j_h) \in \bar{T}$, respectively $\bar{T}^*$.

Then $\bar{T}, \bar{T}^* \in \mathcal{T}^v_{m-1}(\{k_i\}_{i \neq h})$, $\bar{T} \cap \bar{T}^*$ contains exactly $|T \cap T^*| - (k_h + 1)$ edges in common and, in particular, all edges $(\{C^*_i\}_{i \neq h})$. By induction we know that one step obtains a neighbor $\bar{T}'$ of $\bar{T}$ which has one more edge in common with $\bar{T}^*$. Suppose $(i, j) \in \bar{T}^* \sim \bar{T}$ is the edge which is so adjoined to $\bar{T}$, and $(i, \ell) \in \bar{T} \sim \bar{T}^*$ the edge which is dropped. Either (a) $j \neq j_h, \ell \neq j_h$ or (b) $j = j_h, \ell \neq j_h$ or (c) $j \neq j_h, \ell = j_h$ (clearly it is impossible for $j = j_h$ and $\ell = j_h$).

(a) $(i, j) \in T^* \sim T$, $(i, \ell) \in T \sim T^*$ and $T' = T \cup \{(i, j)\} \sim \{(i, \ell)\}$ is a neighbor of T which is "better", that is, coincides with T^* in one more edge.

(b) There exists a unique edge $(i, j) \in T^* \sim T$, $j \in J_h, (i, \ell) \in T \sim T^*$ and $T' = T \cup \{(i, j)\} \sim \{(i, \ell)\}$ is better.

(c) There exists a unique edge $(i, \ell) \in T \sim T^*$, $\ell \in J_h$, $(i, j) \in T^* \sim T$ and $T' = T \cup \{(i, j)\} \sim \{(i, \ell)\}$ is better.

This completes the proof.

Given any polytope P, call the *distance* between any pair of extreme points the number of extreme edges in a shortest path linking the extreme points. Call the *diameter* of P the greatest distance between any pair of extreme points.

Theorem 1.8 (Hirsch conjecture). *The diameter of* $P_{m,n}(\{k_i m + 1\}; \{m\})$ *is* $\sum k_i + m = \bar{k} + m = m + n - 1$ *for* $m, n > 2$; *is* n *for* $m = 2$; *and is* m *for* $n = 2$ (*or* $\bar{k} = 1$).

Proof. Theorem 1.1, together with Lemmas 1.6 and 1.7, establish a diameter of at most $m + n - 1$ if $k_i \geq 1$. Suppose, then, that $k_i = 0$ for $i \in I_0 \subset I$. This means that $T \in \mathcal{T}^v_m$ has valency 1 for nodes $i \in I_0$. Thus, given any two trees T and T^*, at most $|I_0|$ steps obtains a pair T', T^* having the edges incident at nodes $i \in I_0$ in common. Since no edge (i,h) subsequently adjoined to T' or its successors has $i \in I_0$, these can be dropped from further consideration. If $m = 2$ or $n = 2$, then any pair of trees must share at least one edge.

2. $P_{m,n}(\{(k_i + 1)m - 1\}; \{m\})$

Consider the class of polytopes $P_{m,n}(\{(k_i + 1)m - 1\}; \{m\})$, where $k_i \geqslant 0$ and integer. Then n is defined via $mn = \sum_1^m (k_i m + m - 1)$ or, taking $\bar{k} = \sum_1^m k_i (\geqslant 0)$, $n = \bar{k} + m - 1$. A vertex x of P has, as its graph, $G(x)$ which must be a spanning tree of $C(I,J)$, $|I| = m$, $|J| = \bar{k} + m - 1 = n$, and hence contains exactly $\bar{k} + 2m - 2$ edges.

Define the particular family of spanning trees of $C(I,J)$, $\mathcal{T}_m^e(\{k_i\})$, called trees of *source-constant end-sinks*, to be trees in which each source $i \in I$ has precisely $k_i \geqslant 0$ adjacent end-sinks.

Theorem 2.1. *The trees $\mathcal{T}_m^e(\{k_i\})$ are in one-one correspondence with the extreme points of $P_{m,n}(\{(k_i + 1)m - 1\}; \{m\})$, $n = \bar{k} + m - 1$, $\bar{k} = \sum_1^m k_i$.*

Proof. First, consider any vertex x of P. x_{ij} can be equal to m for at most k_i indices j, so $G(x)$ can have at most k_i end-sinks. Remove all end-sinks from $G(x)$ to obtain G', a tree on m sources and n' sinks, each sink having G'-valency at least 2. Then G' has at least $2n'$ edges, so $m + n' - 1 \geqslant 2n'$ or $m - 1 \geqslant n'$. But $n' \geqslant n - \sum k_i = m - 1$, so $n' = m - 1$ and each source is a $G(x)$-neighbor of exactly k_i end-sinks.

Second, consider any $T \in \mathcal{T}^e$. If $(i, j) \in T$ with j an end-sink of T, then let $x_{ij} = m$ and remove (i,j) from T to obtain a tree T' on m sources and $m - 1$ sinks. T' is a tree of constant source valency with $k_i = 1$, all i, $T' \in \mathcal{T}_{m-1}^v(\{1\})$, where here the $m - 1$ left over $j \in J$ are taken as sources and the m nodes $i \in I$ as sinks. Thus Theorem 1.1 applies and the proof is complete.

It should be noted that for $T \in \mathcal{T}_m^e(\{k_i\})$, every $j \in J$ is of T-valency exactly 1 or exactly 2. Thus, in the sequel, the notations $J_1(T)$ and $J_2(T)$ will be used, respectively, to represent the set of end-sinks (of cardinality $\bar{k}$), and the set of sinks of valency 2 (of cardinality $m - 1$).

Theorem 2.2. $|\mathcal{T}_m^e(\{k_i\})| = (\bar{k} + m - 1)! m^{m-2} / \prod_1^m k_i!$, *where* $\bar{k} = \sum_1^m k_i$.

Proof. There are $\binom{\bar{k}+m-1}{k_1}$ ways of assigning k_1 end-sinks to source-node 1, $\binom{\bar{k}+m-1-k_1}{k_2}$ remaining ways of assigning k_2 end-sinks to source-node 2, $\ldots$ the left-over $m - 1$ sink-nodes J_2 each have valency 2 and their edges form a spanning tree of $C(I,J_2)$ of which, by Theorem 1.4, there are precisely $(m - 1)! m^{m-2}$.

Two trees of $\mathcal{T}^e$ are neighbors if and only if the union of their edges contains exactly one cycle. So, given $T \in \mathcal{T}^e$, a neighbor is obtained by adjoining some edge $(i, j) \notin T$ and dropping some edge in the cycle thus formed to obtain $T' \in \mathcal{T}^e$. If (i, j) is adjoined with $j \in J_2(T)$, then the edge $(k, j) \in T$ on the cycle formed must be dropped (see Fig. 2). If (i, j) is adjoined with $j \in J_1(T)$, where $(k, j) \in T$, then the edge (k,h) on the cycle formed must be dropped (see Fig. 2).

$$x = (x_{ij}) = \begin{bmatrix} 3 & 3 & 2 & \\ & 1 & 1 & \\ & & 2 & 3 \end{bmatrix} \; ; \quad x \in P_{3,5}(8, 2, 5; \{3\});$$

$G(x) = T; \qquad T, T' \text{ neighbors}; \qquad T, T'' \text{ neighbors}; \qquad |\mathcal{T}_3^e(2, 0, 1)| = 180$

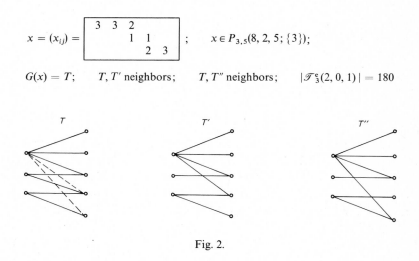

Fig. 2.

As for the first class of trees considered, the straight-forward greedy approach does not work. The same example as given in Fig. 1 provides the counterexample since $\mathcal{T}_2^v(1,1,1) = \mathcal{T}_3^e(0,0)$. Again, an "intermediate" objective in going from T to T^* must first be satisfied.

A *full set of claws* $\{C_i\}$ for $C(I, J)$, $|I| = m$, $|J| = \bar{k} + m - 1$ is a set of independent claws with $|C_i| = k_i + 1$ for all $i \in I$ except one, say h, for which $|C_h| = k_h$. Thus $\sum_i |C_i| = \bar{k} + m - 1$. Any $T \in \mathcal{T}^e$ possesses a full set of claws: if $(i, j) \in T$ and $j \in J_1(T)$ take $(i, j) \in C_i$, then choose as root $r \in I$ some node for which $k_r > 0$ (or, if $\bar{k} = 0$, make an arbitrary choice $r \in I$) and take a set of rooted claws on $C(J_2, I)$ (see Lemma 1.2).

Lemma 2.3. *Given $T \in \mathcal{T}_m^e(\{k_i\})$ and any full set of claws $C^* = \{C_i^*\}$, some T^* containing $\{C_i^*\}$ can be obtained in $\sum (k_i + 1) - 1 - \sum |T \cap C_i^*|$ $\leqslant \bar{k} + m - 1$ steps.*

Proof. The proof is by induction on $(m,\bar{k})$. It is only necessary to show that it is always possible to introduce an edge of C^* into T to obtain a neighbor T' with no edge of C^* dropping.

The lemma is clearly true for $\mathcal{T}^e_m(\{0\}) = \mathcal{T}^v_{m-1}(\{1\})$, for $\mathcal{T}^e_1(\{k_i\})$, and for $\mathcal{T}^e_2(\{k_i\}) = \mathcal{T}^v_2(\{k_i + 1\})$ by Lemma 1.6.

Consider, then, the trees $\mathcal{T}^e_m(\{k_i\})$, with T, C^* given. Suppose $j \in J_2(T)$. Then, if $(k, j) \in C^*$, $(k, j) \notin T$, adjoining (k, j), hence deleting an edge of T incident to j, must clearly result in a neighbor T' better by 1. So it may be assumed that if $j \in J_2(T)$, then its C^*-edge (k, j) is also in T. Thus, $|T \cap C^*| = \sum |T \cap C^*_i| \geq m - 1$.

Suppose that for every $i \in I$ there exists an edge $(i, j) \in T$ with $(i, j) \notin C^*$ and $j \in J_2(T)$. Then there exist at least m such edges, each (by the above assumption) paired with an edge of C^*, implying $|J_2(T)| \geq m$, which is a contradiction. Therefore, *either* (a) some i has edges $(i, j) \in T$, $(i, j) \notin C^*$ and for all such edges $j \in J_1(T)$ *or* (b) some i has all adjacent edges $(i, j) \in T \cap C^*$.

(a) $(k, j) \in C^*$ for some $k \in I$. Adjoin (k, j) to T. No edge of C^* can drop since, in fact, for some edge $(i, h) \in T \cap C^*_i$, h becomes an end-sink (to replace (i, j)) in T'.

(b) Suppose the T-valency of i, which in this case is equal to $|C^*_i|$, is greater than 1. Then $k_i \geq 1$ and, hence, for some edge $(i, j) \in C^* \cap T$, $j \in J_1(T)$. Drop (i, j) from T and from C^* to obtain a pair $\bar{T}$, a tree, and $\bar{C}^*$, a full set of claws, for $\mathcal{T}^e_m(k_1, \ldots, k_i - 1, \ldots, k_m)$. By induction, improvement is always possible in this family of trees, and since such steps never involve forming a cycle including the node $j \in J$, these same steps establish improvement in $\mathcal{T}^e_m(\{k_i\})$.

If, on the other hand, the T-valency of i, equal to $|C^*_i|$, is 1, then $k_i = 0$. Drop the unique edge (i, j) incident at i, and the unique second edge incident at j, say (h, j), from T to obtain $\bar{T}$, and drop (i, j) from C^* to obtain $\bar{C}^*$. Then $\bar{T}$ is a tree and $\bar{C}^*$ a full set of claws of $\mathcal{T}^e_{m-1}(\{k_j\}_{j \neq i})$. By induction, improvement is always possible in this family of trees, and since such steps never involve a cycle including the node $j \in J$, these same steps establish improvement in $\mathcal{T}^e_m(\{k_i\})$. This completes the proof.

Lemma 2.4. *If $T, T^* \in \mathcal{T}^e_m(\{k_i\})$ have a common full set of claws, then T^* can be obtained from T in $|T^*| - |T \cap T^*| \leq m - 1$ steps.*

Proof. The proof is again given by induction on m and $\bar{k}$. The lemma is true for $\mathcal{T}_m^e(\{0\}) = \mathcal{T}_{m-1}^v(\{1\})$ and $\mathcal{T}_2^e(\{k_i\}) = \mathcal{T}_2^v(\{k_i + 1\})$ by Lemma 1.7.

Now, note the following facts. If $(i, j) \in T, j \in J_1(T)$, then $(i, j) \in T^*$ and, symmetrically, if $(i, j) \in T^*$, $j \in J_1(T^*)$, then $(i, j) \in T$, since T and T^* contain a common full set of claws $C^* = \{C_i^*\}$. Also, every node $i \in I$ in any tree of $\mathcal{T}_m^e(\{k_i\})$ must have valency at least $k_i + 1$, and at least two nodes $i \in I$ must have valency equal to $k_i + 1$ (otherwise the total number of edges of the tree would be too great). From these two facts it can be concluded that there must exist some node $i \in I$ with $|C_i^*| = k_i + 1 = T$-valency of i.

It is convenient to distinguish three possible cases and show that in each "improvement" is possible, in the usual sense that a neighbor of T can be found having one more edge in common with T^* than T has.

(a) $k_i \geqslant 2$. The T-valency of i is $k_i + 1$, the T^*-valency of i is at least $k_i + 1$, i has k_i end-sinks in T and T^*. There exists an edge $(i, j) \in C^* \subset T \cap T^*$ with $j \in J_1(T) \cap J_1(T^*)$. For otherwise, $(i, j) \in T^*, j \in J_1(T^*)$ implies $(i, j) \in C^* \subset T$ and $j \in J_2(T)$. But this means i in T has at least k_i adjacent edges (i, j) with $j \in J_2(T)$, together with k_i adjacent end-sinks, hence its valency is at least $k_i + k_i > k_i + 1$, a contradiction. Drop (i, j) from T and T^* to obtain, respectively, $\bar{T}$ and $\bar{T}^*$ in $\mathcal{T}_m^e(k_1, \ldots, k_i - 1, \ldots, k_m)$, $\bar{T}$ and $\bar{T}^*$ containing a common full set of claws. By induction, improvement is always possible in $\mathcal{T}_m^e(k_i, \ldots, k_i - 1, \ldots, k_m)$ and, clearly, such steps can never involve (i, j), so these same steps lead to improvement in the corresponding trees of $\mathcal{T}_m^e(\{k_i\})$.

(b) $k_i = 1$. $|C_i^*| = 2$, T-valency of i is (, say $(i, j_1), (i, j_2)$ are the only edges incident to i in T, with $j_1 \in J_1(T), j_2 \in J_2(T)$. Then, $(i, j_1), (i, j_2)$ also belong to T^*. If $j_1 \in J_1(T^*)$, then the same reasoning used in (a) applies, and improvement is possible. Otherwise, (see Fig. 3) since T^* must have

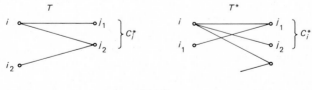

Fig. 3.

some edge $(i, j), j \in J_1(T^*)$, and this edge must belong to C_i^*, hence to T, then $j = j_2$ or $j_2 \in J_1(T^*)$. Thus $j_1 \in J_2(T^*)$ and adjoining (i_1, j_1), the unique second edge incident to j_1 in T^*, to T must result in dropping the unique second edge (i_2, j_2) incident to j_2 in T, and clearly $(i_2, j_2) \notin T^*$, so improvement is achieved.

(c) $k_i = 0$. The T-valency of i is 1, with $(i, j) \in T$, $j \in J_2(T)$, with second edge $(h, j) \in T$. $C_i^* = (i, j)$, $(i, j) \in T^*$, $j \in J_2(T^*)$, with second edge $(h^*, j) \in T^*$.

If $h^* \neq h$, then adjoining (h^*, j) to T must result in dropping (h, j) from T, so improvement is achieved. Otherwise, $h = h^*$, and *either* (c_1) the T^*-valency of i is also 1 or (c_2) the T^*-valency of i is greater than 1.

(c_1) Drop edges (i, j), (h, j) from T and T^* to obtain $\bar{T}$ and $\bar{T}^*$, containing in common a full set of claws and belonging to $\mathcal{T}_{m-1}^e(\{k_j\}_{j \neq i})$. Improvement is possible for trees in this latter class, and clearly such changes are valid for $\mathcal{T}_m^e(\{k_i\})$.

(c_2) Let (see Fig. 4) (i, j), (i, j_1), ..., (i, j_p) be all edges of T^* incident at i. Since C^* is a common full set of claws for T and T^*, then the unique

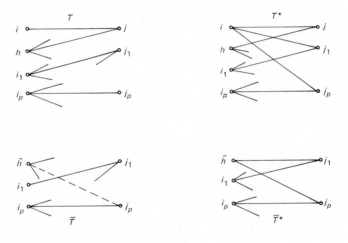

Fig. 4.

second edges incident at j_ℓ ($\ell = 1, ..., p$) in T^* (i_1, j_1), ..., (i_p, j_p) belong to C^* hence to T. If $(h, j_\ell) \in T$ for some ℓ, then adjoining (i, j_ℓ) to T must result in dropping (h, j_ℓ) from T, thus yielding improvement $((h, j_\ell)$ cannot belong to T^* since this would imply the existence of a cycle in T^*). Therefore, assume $(h, j_\ell) \notin T$ for all ℓ. Consider, then, the trees $\bar{T}$, $\bar{T}^*$ obtained, respectively, from T, T^* as follows: $\bar{T}$ is T without nodes i, j and their incident edges (i, j), (h, j), and node h renamed $\bar{h}$; $\bar{T}^*$ is T^*

without nodes i, j, and their incident edges (i, j), (h, j), (i, j_1), $\ldots$, (i, j_p), node h renamed $\bar{h}$ with $(\bar{h}, j_1)$, $\ldots$, $(\bar{h}, j_p)$ adjoined. These trees both belong to $\mathcal{T}^e_{m-1}(\{k_j\}_{j=i})$, contain a common set of claws $\{C_j\}, j \neq i$, and therefore a neighbor $\bar{T}'$ of $\bar{T}$ coinciding with $\bar{T}^*$ in more edges can be obtained. If some edge $(\bar{h}, j_\ell)$ is adjoined to $\bar{T}$, then some edge drops from $\bar{T}$ which does not belong to $\bar{T}^*$ hence is neither an edge directly identified with an edge of T^* nor is any of the edges $(\bar{h}, j_\ell)$. Therefore, taking T instead of $\bar{T}$ and adjoining (i, j_ℓ) to T instead of $(\bar{h}, j_\ell)$ to $\bar{T}$, the same edge of T (or $\bar{T}$) must drop, showing improvement carries over. If, on the other hand, the improvement from $\bar{T}$ to $\bar{T}^*$ involves adjoining some edge $((k, \ell), k \neq \bar{h}$, then the same edge adjoined to T clearly results in the same improvement for T. This completes the proof.

Theorem 2.5 (Hirsch conjecture). *The diameter of* $P_{m,n}(\{(k_i + 1) m - 1\}$; $\{m\})$ *is* $\sum k_i + 2m - 2 = \bar{k} + 2m - 2 = m + n - 1$ *for* $m > 2$ *and is* n *for* $m = 2$.

Proof. Theorem 2.1, together with Lemmas 2.3 and 2.4, establish a diameter of at most $m + n - 1$. Again, if $m = 2$ then any pair of trees must share at least one edge.

3. The Hirsch conjecture status

The Hirsch conjecture is, of course, a long-standing conjecture originally prompted by consideration of the simplex method for linear programming. In linear programming or simplex method terminology it asks: given r independent equations in nonnegative variables—meaning a basis has r variables—is it possible to go from any one feasible basis to any other feasible basis in r pivot steps with each intermediate basis being feasible? Note, that it is obvious at most r pivot steps are necessary if intermediate bases are not required to be feasible. For $P_{m,n}$, $r = m + n - 1$. Geometrically it asks: given a simple convex polyhedron P in p-space defined by q half-spaces, is $q - p$ an upper bound on the diameter of P? For $P_{m,n}$, $p = (m - 1)(n - 1)$, $q = mn$, $q - p = mn - (m - 1)(n - 1) = m + n - 1$.

The conjecture has been established for the polytope associated with the shortest route problem [9]. It has been established for polyhedra arising from Leontief substitution systems [6], a class of polyhedra which includes the shortest route polytope. It has been established (in its linear programming or bases guise) for the assignment polytope [2, 3]. It has also been established for integer-valued extreme points of the set-covering problem $\{x; A x = \bar{1}, x_j = 0,1\}$, where A is a 0–1 matrix and $\bar{1}$ is a column of 1's, a class which includes the assignment polytope [1].

Klee and Walkup [7] elucidate the situation for general polyhedra of dimension less than 6. In particular, they show the Hirsch conjecture for unbounded polyhedra is false for dimensions greater than or equal to 4.

If bets were mandatory, the following would be plausible: with any odds, bet on the truth of the Hirsch conjecture for general transportation polytopes $P_{m,n}$; perhaps without the same confidence, bet on the falsity of the Hirsch conjecture for simple polytopes in general.

References

[1] E. Balas and M.W. Padberg, "On the set-covering problem", *Operations Research*, 20 (1972) 1152–1161.
[2] M.L. Balinski and A. Russakoff, "Some properties of the assignment polytope", *Mathematical Programming* 3 (1972) 257-258.
[3] M.L. Balinski and A. Russakoff, "On the assignment polytope", *SIAM Review*, to appear.
[4] E.D. Bolker, "Transportation polytopes", *Journal of Combinatorial Theory* (B) 13 (1972) 251–262.
[5] G.B. Dantzig, *Linear programming and extensions* (Princeton University Press, Princeton, N.J., 1963) pp. 160, 168.
[6] R.C. Grinold, "The Hirsch conjecture in Leontief substitution systems", *SIAM Journal on Applied Mathematics* 21 (1971) 483-485.
[7] V. Klee and D.W. Walkup, "The d-step conjecture for polyhedra of dimension $d < 6$", *Acta Mathematica* 117 (1967) 53-78.
[8] V. Klee and C. Witzgall, "Facets and vertices of transportation polytopes", in: *Mathematics of the decision sciences*, Vol. I, Eds. G.B. Dantzig and A.F. Veinott, Jr. (Am. Math. Soc., Providence, R.I., 1968) pp. 257–282.
[9] R. Saigal, "A proof of the Hirsch conjecture on the polyhedron of the shortest route problem", *SIAM Journal on Applied Mathematics* 17 (1969) 1232-1238.

Mathematical Programming Study 1 (1974) 59–70. North-Holland Publishing Company

SOLUTION RAYS FOR A CLASS
OF COMPLEMENTARITY PROBLEMS

Richard W. COTTLE

Stanford University, Stanford, Calif., U.S.A.

Received 20 January 1974
Revised manuscript received 18 March 1974

It is shown that when M is a copositive plus matrix, a linear complementarity problem (q, M) has a solution ray at every solution of the problem if and only if q belongs to the boundary of the set of all points p such that the complementarity problem is feasible.

1. Introduction

This work deals with the circumstances under which a linear complementarity problem has a ray of complementary solutions emanating from a given complementary solution. The issues examined here are an outgrowth of the author's earlier papers [1, 2] on monotone solutions of the parametric linear complementarity problem and a recent private communication from Professor G. Maier [8] to whom he is once again indebted for posing a stimulating question.

The introductory paragraphs of many papers on linear complementarity include a passing remark—amounting to nothing more than lip service—to the effect that the linear complementarity problem arises in structural mechanics. Professor Maier's publications in the latter field amply corroborate this often-made assertion. For example, his problem on monotone solutions of the parametric linear complementarity problem [7] is motivated by the desirability of determining (in a particular elastoplastic situation) whether one has regularly progressive yielding and hence can apply the deformation theory or whether one

* Research partially supported by Office of Naval Research contract N-00014-67-A-0112-0011 and National Science Foundation Grant GP 31393X2

has local unloading and hence must use the more cumbersome incremental theories, see [6].

The question Professor Maier recently raised is the following: is it true that if M is symmetric and positive semi-definite, and the linear complementarity problem $(q + \alpha p, M)$:

$$q + \alpha p + M z \geq 0, \tag{1}$$

$$z \geq 0, \tag{2}$$

$$z^{\mathrm{T}}[q + \alpha p + M z] = 0, \tag{3}$$

has a solution $\bar{z} = \bar{z}(\alpha)$ for each α in the interval $[0, \bar{\alpha}]$ but no solution when $\alpha > \bar{\alpha}$, then there is a nonzero $\bar{v}$ such that $\bar{z} + \lambda \bar{v}$ solves $(q + \bar{\alpha} p, M)$ for all $\lambda \geq 0$, and can the symmetry assumption be dropped?

According to Professor Maier, mechanical considerations indicate the answer is yes—at least in the symmetric case.

The response given here is more than just an affirmative answer. It will be shown that not only can the symmetry assumption be dropped, but so can the assumption of the existence of solutions of $(q + \alpha p, M)$ for $\alpha \in [0, \bar{\alpha})$. Moreover, the hypothesis that M is positive semi-definite can be weakened at least to the assumption that M is copositive-plus. Numerical examples suggest the possibility of even greater generality.

The main result, presented in Section 3, characterizes those complementarity problems (r, M) with M copositive-plus for which solution rays exist. It is preceded, in Section 2, by a general lemma giving necessary and sufficient conditions for a solution ray at a particular solution.

2. Necessary and sufficient conditions for solution rays

We shall say the linear complementarity problem (r, M), that is,

$$r + M z \geq 0,$$

$$z \geq 0,$$

$$z^{\mathrm{T}}[r + M z] = 0,$$

has a *solution ray* at $\bar{z}$ with *generator* $\bar{v}$ if and only if $\bar{v} \neq 0$ and $\bar{z} + \lambda \bar{v}$ solves (r, M) for all $\lambda \geq 0$. Notice that such a $\bar{z}$ must itself solve (r, M).

In this section, no assumptions are made about the nature of M. The following lemma is a criterion for a nonzero vector $\bar{v}$ to generate a solution ray for a solvable complementarity problem (r, M) at one of its solutions $\bar{z}$.

Lemma 2.1. *The vector $\bar{v}$ generates a solution ray for (r, M) at $\bar{z}$ if and only if*
  (i) *$\bar{z}$ solves (r, M),*
  (ii) *$\bar{v}$ is a nonzero solution of $(0, M)$,*
  (iii) *$\bar{v}^{\mathrm{T}}(r + M \bar{z}) = 0$,*
  (iv) *$\bar{z}^{\mathrm{T}} M \bar{v} = 0$.*

Proof. If $\bar{v}$ generates a solution ray for (r, M) at $\bar{z}$, then, by definition, $\bar{v} \neq 0$ and $\bar{z} + \lambda \bar{v}$ solves (r, M) for all $\lambda \geq 0$. In particular,

$$r + M (\bar{z} + \lambda \bar{v}) \geq 0, \tag{4}$$

$$\bar{z} + \lambda \bar{v} \geq 0, \tag{5}$$

$$(\bar{z} + \lambda \bar{v})^{\mathrm{T}} [r + M (\bar{z} + \lambda \bar{v})] = 0, \tag{6}$$

for all $\lambda \geq 0$. From (4) and (5), respectively, it follows that

$$M \bar{v} \geq 0, \tag{7}$$

$$\bar{v} \geq 0. \tag{8}$$

Equation (6) says that for all $\lambda \geq 0$

$$\begin{aligned} 0 &= (\bar{z} + \lambda \bar{v})^{\mathrm{T}} [r + M (\bar{z} + \lambda \bar{v})] \\ &= \bar{z}^{\mathrm{T}} r + \bar{z}^{\mathrm{T}} M \bar{z} + \lambda \bar{v}^{\mathrm{T}} [r + (M + M^{\mathrm{T}}) \bar{z}] + \lambda^2 \bar{v}^{\mathrm{T}} M \bar{v}. \end{aligned}$$

Regarded as a quadratic polynomial in λ, the right-hand side of this equation has infinitely many roots, hence all its coefficients must equal zero. Specifically,

$$\bar{v}^{\mathrm{T}} M \bar{v} = 0, \tag{9}$$

$$\bar{v}^{\mathrm{T}}[r + (M + M^{\mathrm{T}})\bar{z}] = 0. \tag{10}$$

Conditions (7), (8) and (9) show that $\bar{v}$ is a solution of $(0, M)$. From (10) we have

$$\bar{v}^{\mathrm{T}}(r + M\bar{z}) + \bar{v}^{\mathrm{T}}M^{\mathrm{T}}\bar{z} = (\bar{v}^{\mathrm{T}})(r + M\bar{z}) + (\bar{z}^{\mathrm{T}})(M\bar{v}) = 0.$$

Since each vector enclosed in parentheses is nonnegative, this equation implies

$$\bar{v}^{\mathrm{T}}(r + M\bar{z}) = 0, \tag{11}$$

$$\bar{z}^{\mathrm{T}}M\bar{v} = 0. \tag{12}$$

These are just parts (iii) and (iv) of the assertion. Conversely, given (i)–(iv), the steps of the arguments above are easily reversed and so the proof is complete.

Remark 2.2. Lemma 2.1 bears a very close resemblance to Lemke's discussion of almost-complementary rays in the context of his pivoting scheme for the linear complementarity problem, see [5, p. 686].

The presence of a solution ray for (r, M) obviously implies the existence of *infinitely many solutions* for the problem. Perhaps the most definitive result on this question published to date is due to Murty [9, p. 73] who showed that (q, M) has a finite number of solutions for all $q \in \mathbf{R}^n$ if and only if M is nondegenerate, i.e., if and only if all the principal minors of M are nonzero. Thus, if (q, M) has a solution ray, M must be degenerate; this is confirmed by the lemma, for $(0, M)$ has a nonzero solution $\bar{v}$. The principal submatrix of M corresponding to the positive components of $\bar{v}$ must be singular.

We can give a bit of geometric flavor to the lemma and at the same time set the stage for Section 3 by discussing the complementary cones relative to M. Credit for emphasizing this concept belongs to Murty, op. cit. Given the $n \times n$ matrix M, we form the $n \times 2n$ matrix $[I, -M]$ and say that the $n \times n$ matrix A is a *complementary matrix* relative to $[I, -M]$ if for each k, $1 \le k \le n$, $A._k$, the kth column of A, equals $I._k$ or $-M._k$. Then for such a matrix,

$$\mathrm{pos}\, A = \{q : q = Ax, x \ge 0\}$$

is called a *complementary cone* relative to M. To solve the linear complementarity problem (q, M) is to obtain q as an element of some complementary cone relative to M. We shall denote by $K(M)$ the union of all the complementary cones relative to M. We observe in passing that for every $n \times n$ matrix M,

$$\mathbf{R}^n_+ \subseteq K(M) \subseteq \mathbf{R}^n.$$

If $K(M) = \mathbf{R}^n$, then M is called a Q-matrix, and if pos $[I, -M] = K(M)$, then M is called a K-matrix. The main unsolved problem in the theory of linear complementarity is to give a complete characterization of Q- and K-matrices.

Let $\bar{z}$ be a solution of (r, M) and define $\bar{w}$ by

$$I\bar{w} - M\bar{z} = r. \tag{13}$$

By performing a principal rearrangement, if necessary, we may assume

$$\bar{w} = \begin{pmatrix} \bar{w}_1 \\ \bar{w}_2 \\ \bar{w}_3 \end{pmatrix} \quad \text{and} \quad \bar{z} = \begin{pmatrix} \bar{z}_1 \\ \bar{z}_2 \\ \bar{z}_3 \end{pmatrix},$$

where $\bar{w}_i$ and $\bar{z}_i$ belong to $\mathbf{R}^{n_i}$ $(i = 1, 2, 3)$, $n_1 + n_2 + n_3 = n$, and

$$\bar{w}_1 > 0, \quad \bar{w}_2 = 0, \quad \bar{w}_3 = 0, \quad \bar{z}_1 = 0, \quad \bar{z}_2 > 0, \quad \bar{z}_3 = 0.$$

Next, suppose $\bar{v}$ generates a solution ray for (r, M) at $\bar{z}$. Write

$$\bar{u} = M\bar{v},$$

and decompose $\bar{u}$ and $\bar{v}$ as above so that

$$\bar{u} = \begin{pmatrix} \bar{u}_1 \\ \bar{u}_2 \\ \bar{u}_3 \end{pmatrix} \quad \text{and} \quad \bar{v} = \begin{pmatrix} \bar{v}_1 \\ \bar{v}_2 \\ \bar{v}_3 \end{pmatrix}.$$

Thus, $\bar{u}_i$ and $\bar{v}_i$ belong to $\mathbf{R}^{n_i}_+$.

Conditions (iii) and (iv) of Lemma 2.1 can be expressed as $\bar{v}^T\bar{w} = 0$ and

$\bar{z}^T\bar{u} = 0$, respectively. Thus we obtain the relations

$$\bar{v}_1^T\bar{w}_1 = 0 \quad \text{and} \quad \bar{z}_2^T\bar{u}_2 = 0$$

which imply

$$\bar{v}_1 = 0 \quad \text{and} \quad \bar{u}_2 = 0.$$

We may even assume that $\bar{w}$ and $\bar{z}$ have positive components where $\bar{u}$ and $\bar{v}$ do. Thus the existence of a solution ray for (r, M) at $\bar{z}$ generated by $\bar{v}$ means that the columns of $[I_1, -M_2, -M_3]$ are positively linearly dependent.

Some simple numerical examples will help to put Lemma 2.1 and Theorem 3.1 into sharper focus.

Example 2.3. Consider the problem (r, M) in which

$$r = \begin{pmatrix} -2 \\ 6 \end{pmatrix} \quad \text{and} \quad M = \begin{pmatrix} 2 & -1 \\ -6 & 3 \end{pmatrix}.$$

A solution of the problem is $\bar{z} = \begin{pmatrix} 1 \\ 0 \end{pmatrix}$, and a generator for a solution ray at $\bar{z}$ is $\bar{v} = \begin{pmatrix} 1 \\ 2 \end{pmatrix}$. In this case, any solution ray generator must be a positive multiple of $\bar{v}$ which also happens to belong to the null space of M. It should be noted that M is neither positive semi-definite nor symmetric. Fig. 1 depicts the complementary cones relative to M.

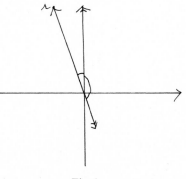

Fig. 1.

Example 2.4. Now consider the problem (r, M) in which

$$r = \begin{pmatrix} -2 \\ 6 \end{pmatrix} \quad \text{and} \quad M = \begin{pmatrix} -1 & 2 \\ 3 & -6 \end{pmatrix}.$$

The only difference between this example and the one above is that their columns have been permuted. Notice that M has negative elements on its main diagonal. Nevertheless, the problem has a solution ray at $\bar{z} = \begin{pmatrix} 0 \\ 1 \end{pmatrix}$ generated by $\bar{v} = \begin{pmatrix} 2 \\ 1 \end{pmatrix}$. See Fig. 2.

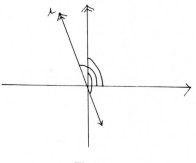

Fig. 2.

Example 2.5. Finally, consider the linear complementarity problem (r, M), where

$$M = \begin{pmatrix} 1 & -3 \\ -3 & 2 \end{pmatrix}$$

and r is any vector in $K(M)$. It will be noted that M is a K-matrix and that, no matter what $r \in K(M)$ is used to form (r, M), there will be no solution ray for (r, M) since $(0, M)$ has only the zero solution. See Fig. 3.

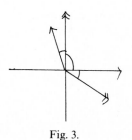

Fig. 3.

3. The main result

In this section, we address the question discussed in the Introduction. We shall restrict our attention here to the class of *copositive-plus* matrices M. These matrices, first identified by Lemke [4], are defined as follows:

  (i) $z^T M z \geq 0$ if $z \geq 0$,
  (ii) $(M + M^T) z = 0$ if $z^T M z = 0$ and $z \geq 0$.

While this class of matrices is not as large as one might hope for in the present context, it is at least large enough to include the situation envisioned in the original formulation of the problem.

In Lemke's path-breaking paper [5] it is shown in Theorem 4 that copositive-plus matrices are K-matrices. For all K-matrices, M, the set $K(M)$ is a convex polyhedral cone with nonempty interior.

If M is a K-matrix and for some $r \in K(M)$ there is a vector p such that $r + \varepsilon p \notin K(M)$ for $\varepsilon > 0$, it follows that r belongs to the *boundary* of $K(M)$. It should be kept in mind that Maier's question (see Section 1) is phrased in terms of a vector r, namely $r = q + \bar{\alpha} p$, belonging to the boundary of $K(M)$, where M is a positive semi-definite (and hence copositive-plus) matrix.

Theorem 3.1. *Let M be a copositive-plus matrix, and let $\bar{z}$ be a solution of the linear complementarity problem (r, M). Then (r, M) has a solution ray at $\bar{z}$ if and only if r belongs to the boundary of $K(M)$.*

Proof. If r belongs to the boundary of $K(M)$, there exists a vector p such that for all $\varepsilon > 0$, $r + \varepsilon p \notin K(M)$. By the separating hyperplane theorem, for each $\varepsilon > 0$ there exists a nonzero vector v such that

$$v \geq 0, \qquad v^T M \leq 0, \qquad v^T(r + \varepsilon p) < 0. \tag{14}$$

The homogeneous inequalities (14) are not satisfied by 0, so we may restrict attention to the nonempty compact convex set

$$V = \{v : v \geq 0, v^T M \leq 0, v^T e = 1\},$$

where $e^T = (1, \ldots, 1)$. This set must contain an element, $\bar{v}$, such that $\bar{v}^T r = 0$. To see this, note that $v^T r \geq 0$ for all $v \in V$, otherwise the inequalities $r + M z \geq 0$, $z \geq 0$ have no solution—but $r \in K(M)$, so they do have a solution, say $\bar{z}$. If $v^T r > 0$ for all v in V, then for $\varepsilon > 0$ sufficiently small it is not possible to find $v \in V$ satisfying (14). Hence V contains an element $\bar{v}$ orthogonal to r. This vector generates a solution ray for (r, M) at $\bar{z}$, for by (14) we have $\bar{v}^T M \bar{v} \leq 0$, whence $\bar{v}^T M \bar{v} = 0$ and $(M + M^T) \bar{v} = 0$ because M is copositive-plus. But now we have $M \bar{v} \geq 0$. This shows $\bar{v}$ is a non-zero solution of $(0, M)$. These facts show further that $(\bar{z} + \lambda \bar{v})^T [r + M (\bar{z} + \lambda \bar{v})] = \bar{z}^T r + \bar{z}^T M \bar{z} + \lambda \bar{v}^T [r + (M + M^T) \bar{v}] + \lambda^2 \bar{v}^T M \bar{v} = 0$ for all λ. Hence $\bar{v}$ generates a solution ray for (r, M) at $\bar{z}$.

To prove the converse, suppose $\bar{v}$ generates a solution ray for (r, M) at $\bar{z}$. If r does not belong to the boundary of $K(M)$, then for all $p \in \mathbf{R}^n$ and all sufficiently small $\varepsilon > 0$, $r + \varepsilon p$ belongs to $K(M)$. By Lemma 2.1,

$$0 \neq \bar{v} \geq 0, \qquad M \bar{v} \geq 0, \qquad \bar{v}^T M \bar{v} = 0,$$
$$\bar{v}^T(r + M \bar{z}) = 0, \qquad \bar{z}^T M \bar{v} = 0. \tag{15}$$

Since M is copositive-plus, $(M + M^T) \bar{v} = 0$ and $\bar{v}^T M \leq 0$. Thus $\bar{z}^T(M + M^T) \bar{v} = 0$, so $\bar{z}^T M^T \bar{v} = 0$ by the last part of (15). Transposing this, we get $\bar{v}^T M \bar{z} = 0$ and then, from (15) again, $\bar{v}^T r = 0$. For all $q \in K(M)$, $v \geq 0$ and $v^T M \leq 0$ imply $v^T q \geq 0$. By our assumption, $r + \varepsilon p \in K(M)$ for *all* $p \in \mathbf{R}^n$ and sufficiently small $\varepsilon > 0$. Applying these observations to our vector $\bar{v}$, we have

$$0 \leq \bar{v}^T(r + \varepsilon p) = \varepsilon \bar{v}^T p$$

which is patently absurd for a non-zero vector $\bar{v}$. This completes the proof.

Actually the converse part of this theorem is related to a corollary of Murty's theorem quoted above. It states that if (r, M) has infinitely many solutions, then r belongs to a complementary cone having an empty interior, see [9, p. 75]. However, it is possible for r to belong to two (or more) complementary cones of which one has an interior and the other does not. This is illustrated by the following example

Example 3.2. Let

$$M = \begin{pmatrix} -1 & 1 \\ -1 & 1 \end{pmatrix}.$$

Note that now $K(M)$ is not convex. All nonzero solutions of $(0, M)$ are positive multiples of $\bar{v} = \binom{1}{1}$, and $\bar{u} = M \bar{v} = 0$. We consider three cases.

Case 1. $r^1 = \binom{0}{1}$. This point belongs to the boundary of $K(M)$. The only solution of (r^1, M) is $\bar{z} = \binom{0}{0}$ for which $\bar{w} = \binom{0}{1}$. Thus $\bar{v}^T \bar{w} > 0$, so there is no solution ray at $\bar{z}$.

Case 2. $r^2 = \binom{1}{1}$. This point does not belong to the boundary of $K(M)$. Indeed it belongs to two complementary cones only one of which has an interior. In this case, $\bar{z} = \binom{1}{0}$ is a solution of (r^2, M) for which $\bar{w} = 0$. Hence $\bar{v}^T \bar{w} = \bar{z}^T \bar{u} = 0$. Hence $\bar{v}$ generates a solution ray at $\bar{z}$.

Case 3. $r^3 = \binom{-1}{1}$. This point belongs to the boundary of $K(M)$. The vector $\bar{z} = \binom{0}{1}$ solves (r^3, M) and gives $\bar{w} = 0$. Thus $\bar{v}^T \bar{w} = \bar{z}^T \bar{u} = 0$, so there is a solution ray for (r^3, M) at $\bar{z}$.

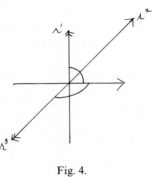

Fig. 4.

Remark 3.3. It is very well known that the necessary and sufficient conditions of optimality for symmetric dual linear programming problems can be stated in terms of the linear complementarity problem. In this instance, M turns out to be skew-symmetric. Hence the theorem has an application to linear programming theory. As a matter of fact, a paper by Goldman and Tucker [4] deals in part with rays of optimal solutions for the primal or the dual individually. Goldman and Tucker establish the relationship between the existence of such optimal rays and "admissible numbers" for associated matrix games. They also give a structure theorem for the set of all optimal solutions of either problem. However, none of these results seem to concern the relationship of the data to boundary of $K(M)$.

It would be of some interest to establish the theorem above for a larger class of (not-necessarily symmetric) matrices—a class which properly includes the copositive-plus matrices. The author has already taken a few tentative steps in this direction. For example, it is easy to show that if M has the property (see Example 2.4)

$$v \geq 0, \qquad v^{\mathrm{T}}M \leq 0 \text{ implies } Mv \geq 0, \tag{16}$$

and r belongs to the boundary of $K(M)$, then (r, M) has a solution ray at any of its solutions. With a little more effort, the converse can be proved if M belongs to the class L_2 defined by B.C. Eaves [3, p. 619]. Ideally, one would like a theorem which completely characterizes the class of matrices for which the result established here is valid.

4. A Tucker retrospective

This paper is affectionately dedicated to Professor A.W. Tucker on the occasion of his retirement from regular duties at Princeton University. Its subject lies very close to one of Professor Tucker's research interests. Unquestionably, this investigation owes much of its existence to his many contributions. It is my pleasure to point out some of the direct and indirect ways these contributions have influenced the present work.

Without wishing to stir up controversy or to indulge in a self-serving exercise, I'd like to comment on how the linear complementarity problem came about. From my recollection, it was Tucker's paper, "Dual systems of homogeneous linear relations" [c] that suggested the idea of expressing symmetric dual quadratic programs in the composite format which lead to the identification of the (linear) complementarity problem—formerly called the fundamental problem. The celebrated Kuhn–Tucker theorem on necessary conditions of optimality played a key role in the early study of the existence of solutions. Later, the development of principal pivoting algorithms was strongly influenced and advanced by Tucker's work on pivotal algebra [e, f].

It might also be remembered that in 1957 the role of quadratic programming in various physical least-action principles was brought to the attention of the operations research community in Tucker's paper [d]. Such applications include those in structural mechanics.

The contributions [a, b] of Tucker's student T.D. Parsons are a further illustration of his influence on his work. For instance, one finds in [b, p. 567] the first reference to the illusive class K.

If Tucker's male students may be called his "sons", then *their* male students must be his "grandsons". One of these is K.G. Murty (via D. Gale). Murty's work cited in this paper introduces the class Q and the important notion of complementary spanning cones.

Although I am not one of Al Tucker's own sons, I hope always to be regarded as a good friend of the family.

[a] T.D. Parsons, "A combinatorial approach to convex quadratic programming", Doctoral dissertation, Department of Mathematics, Princeton University, Princeton, N.J. (May 1966).

[b] T.D. Parsons, "Applications of principal pivoting", in: *Proceedings of the Princeton symposium on mathematical programming*, Ed. H.W. Kuhn (Princeton University Press, Princeton, N.J., 1970) pp. 567–581.

[c] A.W. Tucker, "Dual systems of homogeneous relations", in: *Linear inequalities and related systems*, Eds. H.W. Kuhn and A.W. Tucker (Princeton University Press, Princeton, N.J., 1956).

[d] A.W. Tucker, "Linear and nonlinear programming", *Operations Research* 5 (1957) 244–257.

[e] A.W. Tucker, "A combinatorial equivalence of matrices", in: *Proceedings of the symposia in applied mathematics*, Volume 10, Eds. R. Bellman and M. Hall (A.M.S., Providence, R.I., 1960) pp. 129–140.

[f] A.W. Tucker, "Principal pivot transforms of square matrices", *SIAM Review* 5 (1963) 305.

References

[1] R.W. Cottle, "Monotone solutions of the parametric linear complementarity problems", *Mathematical Programming* 3 (1972) 210–224.

[2] R.W. Cottle, "Solution of 'Problem 72–7, A parametric linear complementarity problem' by G. Maier", *SIAM Review* 15 (1973) 381-384.

[3] B.C. Eaves, "The linear complementarity problem", *Management Science* 17 (1971) 612–634.

[4] A.J. Goldman and A.W. Tucker, "Theory of linear programming", in: *Linear inequalities and related systems*, Annals of Mathematics Studies, No. 38, Eds. H.W. Kuhn and A.W. Tucker (Princeton University Press, Princeton, N.J., 1956) pp. 53–97.

[5] C.E. Lemke, "Bimatrix equilibrium points and mathematical programming", *Management Science* 11 (1965) 681–689.

[6] G. Maier, "A matrix structural theory of piecewise linear elastoplasticity with interacting yield planes", *Meccanica* 5 (1970) 45–66.

[7] G. Maier, "Problem 72–7, A parametric linear complementarity problem", *SIAM Review* 14 (1972) 346–365.

[8] G. Maier, Private communication (18 October, 1973).

[9] K.G. Murty, "On the number of solutions of the complementarity problem and spanning properties of complementary cones", *Linear Algebra and Its Applications* 5 (1972) 65–108.

Mathematical Programming Study 1 (1974) 71–95. North-Holland Publishing Company

ON FOURIER'S ANALYSIS OF LINEAR
INEQUALITY SYSTEMS* **

R.J. DUFFIN

Carnegie-Mellon University, Pittsburgh, Pa., U.S.A.

Received 23 August 1973
Revised manuscript received 19 February 1974

Fourier treated a system of linear inequalities by a method of elimination of variables. This method can be used to derive the duality theory of linear programming. Perhaps this furnishes the quickest proof both for finite and infinite linear programs. For numerical evaluation of a linear program, Fourier's procedure is very cumbersome because a variable is eliminated by adding each pair of inequalities having coefficients of opposite sign. This introduces many redundant inequalities. However, modifications are possible which reduce the number of redundant inequalities generated. With these modifications the method of Fourier becomes a practical computational algorithm for a class of parametric linear programs.

1. Concepts and examples

Fourier proposed to treat a system of linear inequalities by elimination of variables. Thus a variable, say x_h, may be eliminated from a pair of inequalities if the coefficients of the variable have opposite sign. Then a new system of linear inequalities can be formed comprising the inequalities formed from all such pairs together with the inequalities which did not contain x_h. This process may be termed *pairwise elimination*.

The simplicity and utility of pairwise elimination can be understood from a simple example.

* Dedicated to Albert Tucker who has made such instructive use of tableaus simultaneously depicting primal and dual programs. This is the inspiration for the tableau of the present paper which simultaneously depicts Fourier elimination and its dual, Dines elimination.

** Prepared under Research Grant DA-AROD-31-124-71-G17, Army Research Office, Durham.

Example 1.   Find the range of the function $u = x + y$ when x and y are constrained by the following system of inequalities:

$$2x + y \geq 20, \tag{1.1}$$

$$x - 2y \geq 10, \tag{1.2}$$

$$-3x + y \geq -60, \tag{1.3}$$

$$-3x - y \geq -80. \tag{1.4}$$

Solution.   Use the equality $u = x + y$ to eliminate y. Then

$$x + u \geq 20, \tag{1.1'}$$

$$3x - 2u \geq 10, \tag{1.2'}$$

$$-4x + u \geq -60, \tag{1.3'}$$

$$-2x - u \geq -80. \tag{1.4'}$$

Positive multiples of inequalities can be added, so

$$5u \geq 20, \quad 4(1') + (3'), \tag{1.1''}$$

$$-5u \geq -140, \quad 3(3') + 4(2'), \tag{1.2''}$$

$$u \geq -40, \quad 2(1') + (4'), \tag{1.3''}$$

$$-7u \geq -220, \quad 2(2') + 3(4'). \tag{1.4''}$$

These are equivalent to the single relation

$$4 \leq u \leq 28.$$

Can $u = 28$? If so, (1.2'') must be an equation. Also the relations (1.2') and (1.3') which generated (1.2'') must be equations. Thus (1.2') gives $x = 22$. Then (1.2) gives $y = 6$. The values $x = 22$ and $y = 6$ are seen to satisfy all the inequalities. This solves the linear program—maximize u.

Similarly it can be shown that $u = 4$, $x = 16$ and $y = -12$ solves the linear program—minimize u. More generally, the following theorem holds:

Theorem A. *Any linear program can be solved by pairwise elimination of variables followed by back-substitution.*

For numerical work no a priori proof of this statement is needed because the calculations (as above) justify it in each case.

Since Fourier's time, various treatments of the pairwise elimination method have been given [1–16]. The drive of the present paper is treating parametric linear programs. Thus of concern is a system of linear inequalities S having the following canonical form:

$$S \qquad \sum_{j=1}^{m} a_{ij}x_j \geq \lambda_i \qquad i = 1, \ldots, n.$$

Here the coefficients a_{ij} are decimals, the x_j are variables, and the λ_i are parameters.

The pairwise elimination process is applied to eliminate a variable, say x_h, from S and to obtain a new system of linear inequalities S_h. The system S_h is termed an *eliminant* because x_h does not appear. Then by the same process a second variable, say x_k, is eliminated from S_h to obtain a second eliminant S_{hk}. Conceivably an eliminant is empty. For example, S_h is empty if $a_{ih} > 0$, $i = 1, \ldots, n$.

When all the variables x_1, x_2, $\ldots$, x_m have been eliminated, the resulting system is termed a *solvent system* Λ and is of the form

$$\Lambda \qquad 0 \geq \sum_{i=1}^{n} Y_{ki}\lambda_i \qquad k = 1, \ldots, L.$$

The only operations used in deriving Λ from S are multiplication by positive numbers and addition. Then a little thought shows that the coefficients Y_{ki} are decimals satisfying:

$$Y_{ki} \geq 0 \qquad positivity,$$

$$\sum_{i=1}^{n} Y_{ki}a_{ij} = 0 \qquad orthogonality,$$

and these relations are to hold for $i = 1, \ldots, n$ and $j = 1, \ldots, m$ and $k = 1, \ldots, L$. The array $\{Y_{ki}\}$ is termed a *solvent matrix*. The terminology is justified by the following theorem:

Theorem B. *A system of linear inequalities has at least one solution* x_1, *..., x_m if and only if the solvent system is satisfied.*

A computational difficulty with pairwise elimination is that the number of inequalities in the eliminant system may be very large. Thus the first eliminant could have $(n/2)^2$ inequalities. The final eliminant could have, in an unfavorable case, $L = 4(n/4)^{2^m}$ inequalities. This number may be enormous, even for small problems. To alleviate this difficulty, two modifications are introduced:

(a) A rule for selection of an optimum order of elimination of variables (stepwise optimum);

(b) A rule for deletion of certain redundant inequalities which result from the nature of pairwise elimination.

These rules will be explained in detail in the solution of the following special examples.

Example 2. Determine whether the following system is consistent:

$$S \qquad \begin{aligned} -5x_1 - x_2 - 3x_3 + x_4 &\geq \lambda_1, \\ 2x_1 - x_2 - x_3 + x_4 &\geq \lambda_2, \\ -2x_1 + 2x_2 + 4x_3 - 2x_4 &\geq \lambda_3, \\ 3x_1 + x_2 + 2x_3 - x_4 &\geq \lambda_4, \\ x_1 - x_3 &\geq \lambda_5, \\ -6x_1 + x_2 + 5x_3 + x_4 &\geq \lambda_6. \end{aligned}$$

Solution. Let the *expansion number* E_h of a variable x_h be given by the formula

$$\begin{aligned} E_h &= p_h q_h + r_h \quad \text{if } p_h + q_h > 0, \\ E_h &= 0 \qquad\qquad \text{if } p_h + q_h = 0. \end{aligned}$$

Here p_h, q_h and r_h are the number of positive, negative, and zero coefficients of x_h, respectively.

Rule (a). *Eliminate the variable with minimum positive expansion number.*

In the present system, $E_1 = 3 \times 3 = 9$, $E_2 = 3 \times 2 + 1 = 7$, $E_3 = 4 \times 2 = 8$, $E_4 = 3 \times 2 + 1 = 7$. Thus x_2 or x_4 should be eliminated first. Choose x_2 and system S_2 will have seven inequalities. It is instructive now to present the eliminant systems in a tableau as shown in Fig. 1.

	x_1	x_2	x_3	x_4	λ_1	λ_2	λ_3	λ_4	λ_5	λ_6	pairing	
S	−5	−1	−3	1	1						1	
	2	−1	−1	1		1					2	
	−2	2	4	−2			1				3	
	3	1	2	−1				1			4	
	1		−1						1		5	
	−6	1	5	1						1	6	
S_2 S_{24}	−11		2	2	1					1	7 1 6	
	−4		4	2		1				1	8 2 6	
	−2		−1		1			1			9 1 4	
	5		1			1		1			10 2 4	
	−12		−2		2		1				11 1 3	
	2		2			2	1				12 2 3	
	1		−1							1	13 5	
S_{243} S_{243}^{*}	−2				2	2	1	2			14 9 12	*
	−2				2	2	1	2			15 10 11	*
	−20				4	4	4				16 11 12	
	3				1	1		2			17 9 10	
	4					2	1		2		18 12 13	
	6					1			1	1	19 10 13	
S_{2431}^{+} Λ^{*}					24	44	24	20	20		20 16 19	*
					32	32	12	40			21 16 17	
					16	56	36		40		22 16 18	

Fig. 1. Tableau of eliminant systems.

The tableau makes evident how the eliminant S_2 is formed. Thus by observing the coefficients of the parameters, it is seen that the first row of S_2 is a combination of the first and sixth rows of S. The second row of S_2 is a combination of the second and sixth rows of S. All such pairings are recorded in the last columns of the tableau.

In S_2, we have $p_4 = 2$, $q_4 = 0$, and $r_4 = 5$. It is clear that the minimum expansion number is $E_4 = 5$ and that x_4 is to be eliminated. In the tableau, S_{24} is derived by deleting the first two rows of S_2.

Next, a variable is to be eliminated from S_{24}. Thus $E_1 = 3 \times 2$ and $E_3 = 2 \times 3$, so there is no preference between x_1 and x_3. Eliminating x_3 gives the system S_{243} with six rows. However, some of the rows of S_{243} are redundant and may be deleted.

Rule (b). *After t variables have been actively eliminated, delete any inequality with t + 2 or more parametric terms.*

A variable x_h is said to be *actively* eliminated if $p_h q_h > 0$; it is said to be *passively* eliminated if $p_h q_h = 0$. Consider S_{243} in the light of this rule. Variables x_2 and x_3 have been actively eliminated, but x_4 has been passively eliminated, so $t = 2$. The inequalities of S_{243} marked with an asterisk have $t + 2 = 4$ entries in the parameter columns. Hence the two starred rows are to be deleted, leaving the system S_{243}^*. This latter system is termed a *refined eliminant*.

Finally, pairwise elimination of x_1 yields the system of three rows designated as S_{2431}^+. Now $t = 3$ and the first row of S_{2431}^+ has five parametric entries and is deleted. Thus the refined solvent system is

$$\Lambda^* \qquad \begin{aligned} 0 &\geq 32\lambda_1 + 32\lambda_2 + 12\lambda_3 + 40\lambda_4, \\ 0 &\geq 16\lambda_1 + 56\lambda_2 + 36\lambda_3 + 40\lambda_5. \end{aligned}$$

The next example shows how the solvent matrix applies to a parametric linear program.

Example 3. Find the minimum value M of a variable u subject to the constraints:

$$\begin{aligned} -5x_1 - x_2 - 3x_3 + x_4 &\geq \gamma_1 - u, \\ 2x_1 - x_2 - x_3 + x_4 &\geq \gamma_2 - u, \\ -2x_1 + 2x_2 + 4x_3 - 2x_4 &\geq \gamma_3, \\ 3x_1 + x_2 + 2x_3 - x_4 &\geq \gamma_4, \\ x_1 - x_3 &\geq \gamma_5, \\ -6x_1 + x_2 + 5x_3 + x_4 &\geq \gamma_6. \end{aligned}$$

Solution. Letting $\gamma - u = \lambda_1$, $\gamma_2 - u = \lambda_2, \ldots, \gamma_6 = \lambda_6$ reduces the constraints of Example 3 to those of Example 2. Consequently, the solvent system gives immediately:

$$M = \max \{u_1, u_2\},$$

where

$$\begin{aligned} u_1 &= \tfrac{1}{2}\gamma_1 + \tfrac{1}{2}\gamma_2 + \tfrac{3}{16}\gamma_3 + \tfrac{5}{8}\gamma_4, \\ u_2 &= \tfrac{2}{9}\gamma_1 + \tfrac{7}{9}\gamma_2 + \tfrac{1}{2}\gamma_3 + \tfrac{5}{9}\gamma_5. \end{aligned}$$

The next example shows how the solvent matrix solves the corresponding dual program.

Example 4. Let the function

$$v = \gamma_1 y_1 + \gamma_2 y_2 + \gamma_3 y_3 + \gamma_4 y_4 + \gamma_5 y_5 + \gamma_6 y_6$$

be subject to the following constraints:

$$
\begin{aligned}
& y_1 \ge 0,\, y_2 \ge 0,\, \ldots,\, y_6 \ge 0 && \text{\textit{positivity}}, \\
& -5y_1 + 2y_2 - 2y_3 + 3y_4 + y_5 - 6y_6 = 0 \\
& -\ y_1 - \ y_2 + 2y_3 + \ y_4 \qquad\ + \ y_6 = 0 \\
& -3y_1 - \ y_2 + 4y_3 + 2y_4 - y_5 + 5y_6 = 0 \\
& \ \ \ y_1 + \ y_2 - 2y_3 - \ y_4 \qquad\ + \ y_6 = 0 \\
& \ \ \ y_1 + \ y_2 \qquad\qquad\qquad\qquad\ \ = 1 && \text{\textit{normality}}.
\end{aligned}
$$

with the middle four equations labelled *orthogonality*.

Find the maximum value M' of v and corresponding $y_1, \ldots, y_6$.

Solution. It is seen that any row of the solvent matrix Y of Example 3 satisfies the positivity and orthogonality constraints of Example 4. Moreover, dividing the rows of Y by suitable positive constants makes them satisfy the normality constraint. Thus feasible solutions are:

$$
\begin{aligned}
y^1 &= (\tfrac{1}{2}, \tfrac{1}{2}, \tfrac{3}{16}, \tfrac{5}{8}, 0, 0), \\
y^2 &= (\tfrac{2}{9}, \tfrac{7}{9}, \tfrac{1}{2}, 0, \tfrac{5}{9}, 0).
\end{aligned}
$$

However, it is easy to show that if the constraints of the examples both hold, then $u \ge v$. But if u_1 is the solution of Example 3, $u_1 = v(y^1)$. Likewise, if u_2 is the solution, then $u_2 = v(y^2)$. Thus

$$M' = \max\{v(y^1), v(y^2)\}$$

and y^1 or y^2 is the optimal solution.

The next example shows an application of the solvent matrix to geometric programming.

Example 5. Find the minimum of

$$g_0(t) = c_1 t_1^{-5} t_2^{-1} t_3^{-3} t_4^{1} + c_2 t_1^{2} t_2^{-1} t_3^{-1} t_4^{1}$$

subject to the constraints

$$t_1 > 0, \ldots, t_4 > 0,$$
$$g_1(t) = c_3 t_1^{-2} t_2^2 t_3^4 t_4^{-2} + c_4 t_1^3 t_2^1 t_3^2 t_4^{-1} \le 1,$$
$$g_2(t) = c_5 t_1^1 t_2^0 t_3^{-1} t_4^0 + c_6 t_1^{-6} t_2^1 t_3^5 t_4^1 \le 1.$$

Solution. This is a primal geometric program if $c_i > 0$. The corresponding dual program is stated as follows:

Dual geometric program. Find the maximum of

$$V(y) = \prod_{i=1}^{6} \left(\frac{c_i}{y_i} \right)^{y_i} \lambda_1^{\lambda_1} \lambda_2^{\lambda_2}, \qquad c_i > 0,$$

where

$$\lambda_1 = y_3 + y_4, \qquad \lambda_2 = y_5 + y_6$$

and subject to the constraints stated in Example 4.

Let $v = \sum_1^6 \gamma_i y_i + \alpha$ be a hyperplane tangent to $V(y)$ at some initial point. This determines the parameters $\gamma_1, \ldots, \gamma_6, \alpha$. Then the solution of the linear program of Example 4 furnishes a first approximation to the dual geometric program.

The following example gives a modified procedure to treat inequalities with non-negative variables.

Example 6. Determine whether the system of Example 2 is consistent if it is also required that the variables be non-negative.

Solution. This can be treated by adding the four additional constraints

$$x_1 \ge \lambda_7, \qquad x_2 \ge \lambda_8, \qquad x_3 \ge \lambda_9, \qquad x_4 \ge \lambda_{10},$$

where $\lambda_7, \ldots, \lambda_{10}$ are dummy parameters. Then four additional columns must be added to the tableau of Example 2 (Fig. 1). It is not necessary to add new rows to S, because the new inequalities have such simple form. The new inequalities can be accounted for by a slight change of the algorithm. This modification is shown in the tableau given in Fig. 2. The expansion number of x_h is now taken to be

$$E_h = p_h q_h + q_h + r_h \quad \text{if } p_h + q_h > 0,$$
$$E_h = 0 \qquad\qquad\quad \text{if } p_h + q_h = 0.$$

	x_1	x_2	x_3	x_4	λ_1	λ_2	λ_3	λ_4	λ_5	λ_6	λ_7	λ_8	λ_9	λ_{10}
	-5	-1	-3	1	1									
	2	-1	-1	1		1								
S	-2	2	4	-2			1							
	3	1	2	-1				1						
	1		-1						1					
	-6	1	5	1						1				
	-5		-3	1	1						1			
	2		-1	1		1						1		
	-11		2	2	1				1					
S_2	-4		4	2		1			1					
S_{24}	-2		-1		1			1						
	5		1			1		1						
	-12		-2		2		1							
	2		2			2	1							
	1		-1							1				

Fig. 2. Eliminants with non-negative variables.

The expansion numbers of x_1, x_2, x_3, x_4 in S are 12, 9, 10, 9, respectively. Eliminating x_2 gives the system S_2 shown. The first two rows of S_2 result from the eliminations with $x_2 \geq \lambda_8$. The remaining rows of S_2 are identical with those given in Fig. 1. The completion of the tableau gives a Λ^* with eight rows.

Note. If the system matrix elements a_{ij} are integers, then the solvent matrix elements Y_{ki} are also integers. This is clear because only the operations of multiplication and addition are used in the formation of a tableau.

The following rule is appropriate for a system containing a mixture of equalities and inequalities.

Rule (1). *In a mixed system solve one of the equations for a variable and substitute in the other relations, etc. After all variables have been so eliminated from the equations, apply pairwise elimination to the inequalities.*

Note that rule (1) was used in the solution of Example 1. An alternative procedure to treat a mixed system is:

Rule (2). *Replace each equation by an inequality. Then adjoin the inequality obtained by adding the negative sum of the equations.*

2. The pairwise elimination lemma

It is convenient to employ functional notation in giving a proof of Fourier's method. Thus let S be the system of linear inequalities

$$F_i(x_1, x_2, \ldots, x_m) \geq 0 \qquad i = 1, \ldots, n. \tag{2.1}$$

Here F_i is the affine function $\sum_1^m a_{ij} x_j - \lambda_i$.

To eliminate the variable x_1, partition the integers $1, \ldots, n$ into sets P, Q and R by the following rules:

$$\begin{aligned} i \in P \quad &\text{if } a_{i1} > 0, \\ i \in Q \quad &\text{if } a_{i1} < 0, \\ i \in R \quad &\text{if } a_{i1} = 0. \end{aligned}$$

Then the system of inequalities S is expressible in the following partitioned form:

$$a_{p1}[x_1 + f_p(x_2, \ldots, x_m)] \geq 0 \qquad\qquad p \in P, \tag{2.1P}$$

$$-a_{q1}[-x_1 + f_q(x_2, \ldots, x_m)] \geq 0 \qquad\qquad q \in Q, \tag{2.1Q}$$

$$f_r(x_2, \ldots, x_m) \geq 0 \qquad\qquad r \in R. \tag{2.1R}$$

Let S_1 be the new system of linear inequalities

$$-a_{q1} a_{p1}[f_p(x_2, \ldots, x_m) + f_q(x_2, \ldots, x_m)] \geq 0, \tag{2.2PQ}$$

$$f_r(x_2, \ldots, x_m) \geq 0. \tag{2.2R}$$

The system S_1 is a formal consequence of system S. S_1 is termed an *eliminant system* because the variable x_1 does not appear.

Let $|S|$ denote the number of elements in a set S. Then it is clear that

$$\begin{aligned} |S| &= |P| + |Q| + |R|, \\ |S_1| &= |P||Q| + |R|. \end{aligned}$$

Thus S_1 is empty if R is empty and either P or Q is empty. For example, S_1 is empty if $a_{i1} > 0$ for all i.

Lemma 1 *Pairwise elimination of the variable x_1 from a system of linear inequalities gives an eliminant system of linear inequalities. Then $\tilde{x}_2, \ldots, \tilde{x}_m$ is a solution of the eliminant system if and only if there is an $\tilde{x}_1$, such that $\tilde{x}_1, \tilde{x}_2, \ldots, \tilde{x}_m$ is a solution of the original system.*

Proof. If $\tilde{x}_1, \tilde{x}_2, \ldots, \tilde{x}_m$ satisfies the original system S, then, of course, the inequalities (2.1P) and (2.1Q) are valid. Pairing the inequalities (2.1P) and (2.1Q) and adding shows that (2.2PQ) is valid. Moreover the set (2.2R) is the same as the set (2.1R). Hence $\tilde{x}_2, \ldots, \tilde{x}_m$ satisfies the eliminant system S_1. This proves the first part of the lemma.

Next, suppose $\tilde{x}_2, \ldots, \tilde{x}_m$ is a solution of the eliminant system. If P is not empty, let $\tilde{p} \in P$ satisfy

$$f_{\tilde{p}}(\tilde{x}_2, \ldots, \tilde{x}_m) = \min f_p(\tilde{x}_2, \ldots, \tilde{x}_m) \quad \text{for } p \in P. \tag{2.3}$$

Then define $\tilde{x}_1$ as

$$\tilde{x}_1 = -f_{\tilde{p}}(\tilde{x}_2, \ldots, \tilde{x}_m). \tag{2.4}$$

Relations (2.3) and (2.4) give for any $p \in P$

$$\tilde{x}_1 + f_p(\tilde{x}_2, \ldots, \tilde{x}_m) \geq \tilde{x}_1 + f_{\tilde{p}}(\tilde{x}_2, \ldots, \tilde{x}_m) = 0$$

and this shows that inequalities (2.1P) hold. If, in addition, Q is not empty, then (2.4) and (2.2PQ) give for any $q \in Q$

$$-\tilde{x}_1 + f_q(\tilde{x}_2, \ldots, \tilde{x}_m) = f_{\tilde{p}}(\tilde{x}_2, \ldots, \tilde{x}_m) + f_q(\tilde{x}_2, \ldots, \tilde{x}_m) \geq 0$$

and so inequalities (2.1Q) hold. Consequently, the original system is satisfied if P is not empty.

If P is empty but Q is not empty, let $\tilde{q} \in Q$ satisfy

$$f_{\tilde{q}}(\tilde{x}_2, \ldots, \tilde{x}_m) = \min f_q(\tilde{x}_2, \ldots, \tilde{x}_m) \quad \text{for } q \in Q.$$

Then define $\tilde{x}_1$ as

$$\tilde{x}_1 = f_{\tilde{q}}(\tilde{x}_2, \ldots, \tilde{x}_m).$$

Consequently,

$$-\tilde{x}_1 + f_q(\tilde{x}_2, \ldots, \tilde{x}_m) \geq -\tilde{x}_1 + f_{\bar{q}}(\tilde{x}_2, \ldots, \tilde{x}_m) = 0$$

and so (2.1Q) holds.

Finally, if both P and Q are empty, then $\tilde{x}_1$ may be assigned arbitrarily, say $\tilde{x}_1 = 0$. Thus in all cases $\tilde{x}_1$ may be assigned to satisfy (2.1P), (2.1Q), and (2.1R). This is seen to complete the proof of Lemma 1.

The eliminant S_1 may also be expressed in terms of affine functions, say

$$S_1 \qquad G_k(x_2, \ldots, x_m) \geq 0 \qquad k \in K_1.$$

Here K_1 is a suitable index set. Then the pairwise elimination process may be applied to S_1 to eliminate x_2. This yields the second eliminant S_{12},

$$S_{1\,2} \qquad G_k(x_2, \ldots, x_m) \geq 0 \qquad k \in K_1.$$

where K_{12} is another index set distinct from K_1.

Suppose that the variables $x_1, x_2, \cdots, x_m$ are eliminated in order. Then it is a convenient notation to write $S_1 = S(1)$, $S_{12} = S(2)$, $S_{123} = S(3)$, etc. This gives the sequence of eliminant systems

$$S(1), S(2), \ldots, S(d), \ldots, S(m).$$

Conceivably, the variable x_{d+1} has already dropped out of $S(d)$. In that case, $S(d + 1) = S(d)$. In particular, if $S(d)$ is empty, then so also is $S(d + 1)$ and set $Y_{ki} = 0$.

Theorem 1. *Suppose that the variables $x_1, x_2, \ldots, x_d$ are eliminated in order by pairwise elimination from a linear inequality system S. This results in the eliminant system $S(d)$. Then $\tilde{x}_{d+1}, \tilde{x}_{d+2}, \ldots, \tilde{x}_m$ is a solution of $S(d)$ if and only if there exist values $\tilde{x}_1, \ldots, \tilde{x}_d$ such that $\tilde{x}_1, \tilde{x}_d, \ldots, \tilde{x}_m$ is a solution of S.*

Proof. If $\tilde{x}_1, \ldots, \tilde{x}_m$ satisfies S, then Lemma 1 shows that $\tilde{x}_2, \ldots, \tilde{x}_m$ satisfies $S(1)$. Applying Lemma 1 to $S(1)$ shows that $\tilde{x}_3, \ldots, \tilde{x}_m$ satisfies $S(2)$, etc. This is seen to prove the first part of Theorem 1.

Next suppose that $\tilde{x}_{d+1}, \ldots, \tilde{x}_m$ satisfies $S(d)$. Then by Lemma 1 there is an $\tilde{x}_d$ such that $\tilde{x}_d, \ldots, \tilde{x}_m$ satisfies $S(d-1)$. Apply the same type of argument to $S(d-1)$, etc. This is seen to complete the proof in d steps.

Proof of Theorem B. In particular, $S(m) = \Lambda$, the solvent system. The affine functions of $S(m)$ are of the form

$$G_k = \sum_{i=1}^{n} Y_{ki} F_i(x_1, \ldots, x_m) = -\sum_{i=1}^{n} Y_{ki} \lambda_i$$

and we can take $K(m) = 1, 2, \ldots, L$ for some L. This is seen to prove the theorem.

3. Parametric linear programs

The system of linear inequalities S is now to be treated when the parameters λ_i are linear functions of a single variable u. Thus $\lambda_i = \gamma_i - \beta_i u$, where the β_i and the γ_i are parameters. Then a natural problem is to determine the restrictions on the range of u imposed by the inequalities.

Corollary 1. *Let the variable u be constrained by the system of inequalities*

$$\sum_{j=1}^{m} a_{ij} x_j + \beta_i u \geq \gamma_i \qquad i = 1, \ldots, n. \tag{3.1}$$

Let Y be a solvent matrix, and let

$$B_k = \sum_{i=1}^{n} Y_{ki} \beta_i, \qquad C_k = \sum_{i=1}^{n} Y_{ki} \gamma_i.$$

Then the system (3.1) is solvable if and only if u satisfies the system

$$B_k u \geq C_k \qquad k = 1, \ldots, L.$$

If these are consistent, the permissible range of u is defined by:

$$\min u = \max \{C_k/B_k\} \quad \text{for } B_k > 0,$$
$$\max u = \min \{C_k/B_k\} \quad \text{for } B_k < 0.$$

Proof. This is seen to be a corollary of Theorem B.

Any linear program can be expressed in the above form. Thus the statement of Theorem A is a consequence of the constructive proof of Theorem 1.

The following corollary concerns the dual program.

Corollary 2. *If the program of Corollary 1 has a finite minimum M, then M is also the maximum of the function $v = \sum_1^n \gamma_i y_i$ subject to the constraints:*

$$y_i \geq 0 \quad i = 1, \ldots, n \qquad positivity,$$

$$\sum_1^n y_i a_{ij} = 0 \quad j = 1, \ldots, m \qquad orthogonality,$$

$$\sum_1^n y_i \beta_i = 1 \qquad normality.$$

The solvent matrix Y gives feasible solutions:

$$y_i = Y_{ki}/B_k \quad for \ B_k > 0.$$

Moreover, for some k this is an optimal solution.

Proof. The properties of Y prove the feasibility. Then

$$u = \sum_{j=1}^m \sum_{i=1}^n y_i a_{ij} x_j + \sum_{i=1}^n y_i \beta_i u \geq \sum_{i=1}^n y_i \gamma_i = v.$$

Hence $M \geq v$. But this becomes an equality when k is chosen to maximize C_k/B_k, and the proof is complete.

The Farkas Lemma. *Let u and u_i for $i = 1, \ldots, n$ be homogeneous linear functions. Suppose*

$$u_i \geq 0 \quad implies \quad u \geq 0.$$

Then there exist coefficients y_i such that

$$u \equiv \sum_1^n y_i u_i \quad and \quad y_i \geq 0.$$

Proof. Suppose there are $m + 1$ independent variables. By a substitution we can make u an independent variable. Then the u_i are of the form

$$u_i = \sum_{1}^{m} a_{ij}x_j + \beta_i u.$$

Then Corollary 2 with $\gamma_i = 0$ and $M = 0$ proves the identity.

Corollary 2, with suitable qualifications, holds for infinite systems of inequalities. In another study, pairwise elimination was found to give a very simple resolution of such questions of infinite programming.

4. Proof of the symmetric duality theorem

Consider the following standard form of a linear program.

Primal program. Seek the minimum value M of the function

$$u = \sum_{j=1}^{m} b_j x_j$$

subject to the constraints

$$\sum_{j=1}^{m} a_{ij}x_j \geq c_i \qquad i = 1, \ldots, n,$$
$$x_j \geq 0 \qquad j = 1, \ldots, m.$$

Then the dual program has an (essentially) symmetric form.

Dual program. Seek the maximum value M' of the function

$$v = \sum_{i=1}^{n} c_i y_i$$

subject to the constraints

$$\sum_{i=1}^{n} y_i a_{ij} \leq b_j \qquad j = 1, \ldots, m,$$
$$y_i \geq 0 \qquad i = 1, \ldots, n.$$

The duality theorem may be stated as follows.

Theorem C. *The primal program is consistent and has a finite value M if and only if the dual program is consistent and has a finite value M'. Moreover M' = M.*

Proof. Suppose that the primal program has the finite value M. Then the following system of inequalities has a solution

$$S \qquad \sum_{j=1}^{m} a_{ij}x_j \geq c_i \qquad\qquad i = 1, \ldots, n,$$

$$\sum_{j=1}^{m} -b_j x_j \geq -M \qquad i = n + 1,$$

$$x_{i-n-1} \geq 0 \qquad\qquad i = n + 2, \ldots, n + m + 1.$$

Then let A_{ij} be the matrix a_{ij} augmented by the $m + 1$ rows indicated above. Let Y be the solvent matrix of A, so

$$\sum_{i=1}^{n+m+1} Y_{ki}A_{ij} = 0 \qquad j = 1, \ldots, m.$$

The system S is consistent for M but is inconsistent if M is replaced by anything smaller. Thus by Theorem B there must be a row of the solvent matrix, say Y_{hi}, such that

$$\sum_{i=1}^{n+m+1} Y_{hi}\lambda_i = \sum_{i=1}^{n} Y_{hi}c_i - M Y_{h,n+1} = 0, \qquad Y_{h,n+1} \neq 0.$$

Let

$$\tilde{y}_i = Y_{hi}/Y_{h,n+1}$$

so

$$\sum_{i=1}^{n} c_i \tilde{y}_i = M.$$

Moreover, it is clear that $\tilde{y}_{n+2}, \ldots, \tilde{y}_{n+m+1}$ are slack variables in the orthogonality condition. Thus

$$\sum_{i=1}^{n} \tilde{y}_i a_{ij} - b_j \leq 0 \qquad j = 1, \ldots, m.$$

The conclusion is that the dual program is consistent and $M' \geq M$.

It follows directly from the definition of the primal and dual programs that $v(y) \leq u(x)$. Hence $M' = M$. This proves the first part of Theorem C. The second part follows by symmetry.

5. A dual problem posed by Dines

Lloyd Dines rediscovered the Fourier pairwise elimination algorithm and made various extensions and applications. Moreover, he also developed an algorithm to treat the following problem which is dual to the Fourier problem.

Dines Problem. Find the solutions of the system

$$y_i \geq 0 \qquad\qquad i = 1, \ldots, n \qquad (positivity), \qquad\qquad (5.1)$$

$$\sum_{j=1}^{m} y_i a_{ij} = 0 \qquad j = 1, \ldots, m \qquad (orthogonality). \qquad\qquad (5.2)$$

Dines [5] treated this problem by elimination of equations rather than elimination of variables. In fact the number of variables increases. Later Abadie [11] and Dantzig and Eaves [15] extended the Dines algorithm.

To release these dual problems, it is convenient to employ vector notation. Thus the primal problem may be expressed as

$$S \qquad\qquad A\vec{x} \geq \vec{\lambda}.$$

After d eliminations, S becomes

$$S(d) \qquad\qquad Y_d A \vec{x} \geq Y_d \vec{\lambda},$$

where Y_d is the *partial solvent matrix*. Note that the matrix $Y_d A$ has d zero columns. The dual problem can be expressed as

$$T \qquad\qquad A^T \vec{y} = 0, \qquad \vec{y} \geq 0.$$

After d eliminations, T becomes $T(d)$, the *dual eliminant*:

$$T(d) \qquad\qquad A^T Y_d^T \vec{v} = 0, \qquad \vec{v} \geq 0, \qquad \vec{y} = Y_d^T \vec{v}.$$

Note that matrix $A^T Y_d^T$ has d zero rows. Obviously, a solution $\vec{v}$ of $T(d)$ furnishes a solution $\vec{y}$ of T.

The various primal eliminants are manifest in the rows of an elimination tableau such as shown in Fig. 1. Moreover, the various dual eliminants are manifest in the columns of the tableau. This follows from the above vectorial analysis. For example, the eliminant T_{24} dual to S_{24} is

$$
\begin{aligned}
-2v_1 + 5v_2 - 12v_3 + 2v_4 + v_5 &= 0, \\
-v_1 + v_2 - 2v_3 + 2v_4 - v_5 &= 0, \\
y_1 = v_1 + 2v_3, \qquad y_2 &= v_2 + 2v_4, \\
y_3 = v_3 + v_4, \qquad y_4 &= v_1 + v_2, \\
y_5 = v_5, \qquad y_6 &= 0.
\end{aligned}
$$

Theorem D. *The general solution of the Dines problem is*

$$
y_i = \sum_{k=1}^{L} v_k Y_{ki}, \qquad v_k \geq 0, \tag{5.3}
$$

where Y_{ki} is the solvent matrix.

Proof. Let variables $\lambda_1, \ldots, \lambda_n$ be such that

$$
0 \geq \sum_{i=1}^{n} Y_{ki} \lambda_i \qquad k = 1, \ldots, L. \tag{5.4}
$$

Then by Theorem B there exist $x_1, \ldots, x_m$ such that

$$
\sum_{j=1}^{m} a_{ij} x_j \geq \lambda_i \qquad i = 1, \ldots, n. \tag{5.5}
$$

Then the positivity and orthogonality conditions (5.1) and (5.2) give

$$
0 \geq \sum_{i=1}^{n} y_i \lambda_i. \tag{5.6}
$$

In other words, $(5.4) \Rightarrow (5.6)$. But then the Farkas Lemma states that coefficients v_k exist which satisfy (5.3).

6. Removing redundant constraints

A sequence of linear inequality systems

$$S''(1), S''(2), \ldots, S''(m)$$

termed *semi-refined eliminants*, are defined as follows. The variables $x_1, \ldots, x_m$ are to be eliminated in order. Define $S''(1) = S(1)$. Given $S''(d - 1)$, let $S'(d)$ be the pairwise elimination of x_d from $S''(d - 1)$. Any inequality of $S'(d)$ which has been derived from $d + 2$ or more of the original inequalities of S is deleted. Then $S''(d)$ is the system remaining after all such deletions.

Theorem 2. *Theorem 1 remains valid for the semi-refined eliminant systems.*

Proof. Because $S''(1) = S(1)$, it follows that Theorem 1 is true for $d = 1$. Using the inductive method, suppose the theorem true for any index $< d$. Then by virtue of Lemma 1 it follows that Theorem 1 holds for $S'(d)$ if not for $S''(d)$.

Let the variables $x_{d+1}, \ldots, x_m$ be held fixed at arbitrary values. Then the F_i may be regarded as affine functions of the variables $x_1, \ldots, x_d$. Thus any affine function of $S'(d)$ is of the form

$$G = \sum_{i=1}^{n} b_i F_i(x_1, \ldots, x_d), \qquad b_i \geq 0. \tag{6.1}$$

Suppose that $b_i > 0$ for $d + 2$ or more values of i. Any $d + 2$ affine functions of d variables are linearly dependent, so

$$\sum_{i=1}^{n} b_i F_i(x_1, \ldots, x_d) = \sum_{i=1}^{n} c_i F_i(x_1, \ldots, x_d), \tag{6.2}$$

where $|c_i| > 0$ for at most $d + 1$ values of i. However, $b_i \geq 0$, so by a simple argument (due to Caratheodory) it follows that the further conditions $c_i \geq 0$ can be imposed. For example, let us suppose that $e \leq d + 1$ and

$$c_1 > 0, \ldots, c_e > 0, \qquad c_{e+1} = 0, \ldots, c_n = 0. \tag{6.3}$$

Let an inequality system $R \subset S$ be defined as

$$F_1 \geq 0, \quad F_2 \geq 0, \ldots, F_e \geq 0. \tag{6.4}$$

Then form the corresponding semi-refined system of eliminants $R''(1)$, $\ldots$, $R''(d)$. By the algorithmic process it is clear that

$$R''(d) \subset S''(d). \tag{6.5}$$

No inequality can be deleted from $R'(d)$ because R has $d + 1$ inequalities at most. Thus

$$R''(d) = R'(d). \tag{6.6}$$

This means that Theorem 1 applies to $R''(d)$.

Now suppose that system $S''(d)$ is a valid system of inequalities. Then (6.5) shows that $R''(d)$ is valid. Consequently, there exist values of x_1, $\ldots$, x_d which make R a valid system of inequalities.

Since R has a solution, the relations (6.4), (6.3), (6.2), and (6.1) imply that

$$G \geq 0. \tag{6.7}$$

Lemma 2. *The validity of the inequalities $S'(d)$ implies the validity of the inequalities $S'(d)$.*

Proof. This is true because a relation of the form (6.7) holds for every inequality deleted from $S'(d)$.

Lemma 2 completes the induction and proves Theorem 2.

7. Refined pairwise elimination

Suppose that the variable x_h has been eliminated by pairwise elimination. If at least one pair is formed, we say that x_h has been *actively eliminated*. Otherwise x_h is said to be *passively eliminated*.

A sequence of linear inequality systems

$$S^*(1), S^*(2), \ldots, S^*(m),$$

termed *refined eliminants*, are defined as follows. The variables $x_1, \ldots,$ x_m are to be eliminated in order and put $S^*(1) = S(1)$. Given $S^*(d-1)$ let $S^+(d)$ be the pairwise elimination of x_d from $S^*(d-1)$. If there have been t active eliminations, then any inequality of $S^+(d)$ which has been derived from $t+2$ or more of the original inequalities of S is deleted, and this gives the refined eliminant $S^*(d)$.

Theorem 3. *Theorem 1 remains valid for the refined eliminant systems.*

Proof. This theorem would follow from Theorem 2 if all the variables are actively eliminated. This is so because then $t = d$. It would also follow if the variables actively eliminated precede the variables passively eliminated. This is so because a passive elimination does not introduce new inequalities.

Lemma 3. *Let W be a linear inequality system. If in the formation of W_{hk} the variable x_h is passive, then x_h is also passive in the formation of W_{kh}. Moreover, $W_{hk} = W_{kh}$, and if x_k were active, it remains so.*

Proof. Thus the variable x_h appears in W with $p_h q_h = 0$. Suppose $p_h > 0$, then W_h is the subsystem of W obtained by dropping the inequalities in which x_h appears with a positive coefficient. Thus W_k consists of W_{hk} together with inequalities in which x_h appears with a positive coefficient. Hence x_h is passive in W_k and $W_{kh} = W_{hk}$. A similar argument is valid if $q_h > 0$. If $p_h = q_h = 0$, the lemma is trivial. The activity of x_k is obvious.

Consider the application of Lemma 3 to repeated eliminations. For example, in S_{12345}, suppose that x_2 and x_4 were passively eliminated and that x_1, x_3 and x_5 were actively eliminated. Then applying the commutation rule of Lemma 3 gives

$$S_{12345} = S_{12354} = S_{13254} = S_{13524}.$$

It is clear that Lemma 3 applies to refined as well as to regular elimination, so

$$S^*_{12345} = S^*_{13524}.$$

In other words, *if the active eliminations are carried out first, the same eliminants results.* But as noted above, if the active eliminations are carried out first, then Theorem 3 is a consequence of Theorem 2. This completes the proof.

8. Removing all redundancy

Consider two inequalities, say $G \geq 0$ and $H \geq 0$ derived from the original system of inequalities S. Thus

$$G = \sum_1^n b_i F_i, \qquad H = \sum_{-1}^n a_i F_i,$$

where

$$b_i \geq 0, \quad a_i \geq 0, \quad \sum_1^n b_i > 0, \quad \sum_1^n a_i > 0.$$

We shall say that H *dominates* G if

$$a_i > 0 \quad \text{implies} \quad b_i > 0.$$

Rule (c). *In an eliminant system $S(d)$ a dominated inequality $G \geq 0$ is reduntant and may be deleted. Continuing this deletion process leads finally to a system $S^0(d)$ in which no inequality dominates another. Theorem 1 applies to $S^0(d)$.*

Proof. Let $S^0(d)$ be termed a *minimal system.* Suppose that H dominates G and that $H \geq 0$ is also an inequality of the eliminant system $S(d)$. Since "dominance" is a transitive relation it may be supposed that $H \in S^0(d)$. Let $K = G - \theta H = \sum_1^n (b_i - \theta a_i) F_i = \sum_1^n c_i F_i$. We can choose $\theta > 0$ so that $c_i = b_i - \theta a_i \geq 0$ and so that for some i, say $i = s$, $c_s = 0$ but $b_s = 0$. The case that $K = 0$ is trivial. Otherwise we may suppose, without loss of generality, that

$$c_1 > 0, \quad c_2 > 0, \dots, c_e > 0, \quad c_{e+1} = 0, \dots, c_n = 0.$$

Form an inequality system $R \subset S$ defined as

$$R \qquad F_1 \geq 0, \quad F_2 \geq 0, \dots, F_e \geq 0.$$

Let $R(d)$ be the pairwise eliminant system resulting from elimination of the variables $x_1, \ldots, x_d$ from R. Let $S^{(1)}(d)$ be the system $S(d)$ after deletion of the inequality $G \geq 0$. Clearly this inequality is not an inequality of $R(d)$ because R does not involve F_s while G does. (Of course, $R(d)$ might have an equivalent inequality.)

If $S^{(1)}(d)$ is a valid system of inequalities, it follows that $R(d)$ is a valid system of inequalities. Then by virtue of Theorem 1, there exist $x_1, x_2, \ldots, x_d$ to make R valid. Then referring to the definition of K it follows that $K \geq 0$ is a valid inequality. Also $H \geq 0$ if $S^{(1)}(d)$ is valid, so

$$G = K + \theta H \geq 0.$$

Thus the inequality $G \geq 0$ is redundant. This proves the first part of rule (c).

Proceed by an induction indexed by the number of dominated inequalities deleted. Thus delete some other dominated inequality from the system $S^{(1)}(d)$ giving a system $S^{(2)}(d)$. Then try to employ the above argument to show that Theorem 1 applies to $S^{(2)}(d)$. The only trouble with this argument is that the inequality $G \geq 0$ now might be a member of the subsystem $R(d)$. But then the pairwise algorithm shows that $H \geq 0$ is also a member of $R(d)$. Thus G is dominated by H in $R(d)$ and we delete $G \geq 0$ to form $R^{(1)}(d)$. However, only one inequality has been deleted so by what has been proved, Theorem 1 applies to $R^{(1)}(d)$. Hence Theorem 1 applies to $S^{(2)}(d)$.

It is now clear that this inductive process can be continued until there are no dominated inequalities left. This gives the system $S^0(d)$ and the proof is complete.

Lemma 4. *An inequality $G \geq 0$ of an eliminant system is redundant if and only if it is a non-negative combination of the other inequalities. Moreover, G is dominated.*

Proof. The inequality $G \geq 0$ is redundant if it is a consequence of the other inequalities of the eliminant system. Now think of G as a homogeneous linear function of the $m + n$ variables $x_1, \ldots, x_m$ and $\lambda_1, \ldots, \lambda_n$. This observation shows that the lemma is a direct consequence of the Farkas lemma.

Lemma 4 shows that the redundant inequalities determined by

counting rule (b) are dominated. This property is manifest in the tableau given in Fig. 1. In this tableau it is seen that the refined solvent is also a minimal solvent. This might not be true in general.

9. Computer codes

The concept of the minimal eliminant was introduced by Jean Abadie [11]. The concept of the semi-refined eliminant was introduced by David Kohler [13]. Kohler developed a computer code which at each step used semi-refined elimination. He proposed to use minimal elimination, rule (c), only every so often because checking dominance involves much more computer time. Victor Klee [16] used these procedures together with the expansion number criterion, rule (a).

Jeff Buckwalter and I have also developed a computer code to evaluate the solvent matrix Y. Our code permits optional usage of rule (a), rule (b) and rule (c). On Stigler's diet problem, the A matrix has 5 columns and 10 rows. Refined elimination led to Y having 17 rows. The original Fourier algorithm led to Y having 560 rows. Another example with A having 7 columns and 12 rows led to a refined Y of 19 rows. The original algorithm led to Y having 9641 rows.

References

[1] J.B.J. Fourier, "Solution d'une question particulière du calcul des inégalités", in: *Oeuvres* II (Paris, 1890).
[2] L.L. Dines, "Concerning preferential voting", *American Mathematical Monthly* 24 (1917) 321–325.
[3] L.L. Dines, "Systems of linear inequalities", *Annals of Mathematics* 20 (1919) 191–199.
[4] L.L. Dines, "Definite linear dependence", *Annals of Mathematics* 27 (1925) 57–64.
[5] L.L. Dines, "On positive solutions of a system of linear equations", *Annals of Mathematics* 28 (1927) 386–392.
[6] L.L. Dines and N.H. McCoy, "On linear inequalities", *Transactions of the Royal Society of Canada* 27 (1933) 37–70.
[7] T.S. Motzkin, "Beiträge zur Theorie der linearen Ungleichungen", Dissertation, University of Basel (Jerusalem, 1936).
[8] H.W. Kuhn, "Solvability and consistency for linear equations and inequalities", *American Mathematical Monthly* (1956) 217–232.
[9] S.N. Chernikov, "The solution of linear programming problems by elimination of unknowns", *Doklady Akademii Nauk SSSR* 139 (1961) 1314–1317. [Translation in: *Soviet Mathematics Doklady* 2 (1961) 1099–1103.]

[10] S.N. Chernikov, "Contraction of finite systems of linear inequalities", Zhurnal Vychislitel'noi Matematiki i Matematicheskoi Fiziki 5 (1965) 3–20.

[11] J. Abadie, "The dual to Fourier's method for solving linear inequalities", International Symposium on Mathematical Programming, London, 1964.

[12] R.J. Duffin, "An orthogonality theorem of Dines related to moment problems and linear programming", *Journal of Combinatorial Theory* 2 (1967) 1–26.

[13] D.A. Kohler, "Projections of convex polyhedral sets", Operations Research Center Rept. 67–29, University of California, Berkeley, Calif. (1967).

[14] D.A. Kohler, "Translation of a report by Fourier on his work on linear inequalities", *Opsearch* 10 (1973) 38–42.

[15] G.B. Dantzig and B.C. Eaves, "Fourier–Motzkin elimination and its dual", Dept. of Operations Research Rept., Stanford University, Stanford, Calif. (January 1973).

[16] V. Klee, "Vertices of convex polytopes", Lecture Notes, Department of Mathematics, University of Washington, Seattle, Wash. (1973).

Mathematical Programming Study 1 (1974) 96–119. North-Holland Publishing Company

SOLVING PIECEWISE LINEAR CONVEX EQUATIONS*

B. Curtis EAVES

Stanford University, Stanford, Calif., U.S.A.

Received 8 October 1973
Revised manuscript received 1 May 1974

An algorithm is developed for solving $F(x) = y$, where $F: \mathbf{R}^n \to \mathbf{R}^n$ is convex and piecewise linear. The algorithm is based upon complementary pivoting and proceeds by generating paths of solutions to $F(x) = a + bz$.

1. Introduction

Herein we develop an algorithm for solving systems $F(x) = y$ where $F : \mathbf{R}^u \to \mathbf{R}^n$ is the maximum of a finite number of linear functions, which is to say, F is convex and piecewise linear. The algorithm proceeds by generating paths $(x(\cdot), z(\cdot))$ of solutions to systems of form $F(x) = a + bz$.

The general technique employed is an application of the homotopy principle as described in [7]. Namely to compute a solution to a given problem one first devises a deformation from a trivial problem to the given one. Beginning with the solution to the trivial problem, one solves the given problem by following a path, perhaps involving regressions, of solutions through the deformation. In a more concrete sense the algorithm proposed is of the complementary pivot type and is based upon ideas found in [2, 6–8, 9–11, 13, 15, 16, 18–20, 25]; each of these references treat a piecewise linear structure and generate paths therein.

Toward stating our main result, we amplify our terminology. By piecewise linear we really mean continuous and piecewise affine with a finite number of pieces. By a path $(x(\cdot), z(\cdot))$ of solutions to $F(x) = a + bz$ we mean a piecewise linear function $(x(\cdot), z(\cdot)) : J \to \mathbf{R}^{n+1}$ so that

* This research was supported in part by Army Research Office, Durham Contract DAHC-04-71-C-0041, and NSF Grant GP-34559.

$$F(x(t)) = a + b\,z(t)$$

for $t \in J \subseteq [0, +\infty)$. A matrix $H \in \mathbf{R}^{n \times n}$ is defined to be a subgradient of the convex piecewise linear function F at y if

$$F(x) \geq F(y) + H(x - y)$$

for all $x \in \mathbf{R}^n$. We say that convex F is regular at y if all subgradients at y are nonsingular and that F is regular if it is so everywhere. If $F(S) = y$, where S is an infinite set of form $\{x_1 + \theta\,x_2 : \theta \in [0, +\infty)\}$, we say that S is a ray of solutions to $F(x) = y$. Our main result is:

Theorem 1.1. *Let x_0 be a point at which F is regular, let $a = F(x_0)$, and select $b \neq 0$. The algorithm generates a path $(x(\cdot), z(\cdot))$ of solutions to $F(x) = a + b\,z$, where $(x(0), z(0)) = (x_0, 0)$. The path contains at least one of*
  (a) *A solution $x \neq x_0$ of $F(x) = a$.*
  (b) *A ray of solutions to $F(x) = a + b\,z$ for some $z \geq 0$.*
  (c) *A solution to $F(x) = a + b\,z$ for every $z \geq 0$.*

Hence, for example, if it is known that $F(x) = F(x_0)$ has a unique solution and that $F(x) \to \infty$ as $x \to \infty$ (for example, if F is regular), then we can solve $F(x) = y$ for any y by setting $b = y - a$, by observing that the algorithm must terminate with (c), and by observing that $F(x(\bar{t})) = a + b\,z(\bar{t}) = y$, where $z(\bar{t}) = 1$.

This study is an outgrowth of [8] which in turn was motivated by the contention that complementary pivot methods could be used to advantage in the context of dynamic programming; this point of view has yielded some minor successes but has not yet proved to be worthwhile. If F is merely piecewise linear, that is, not convex, results very close to the theorem are still obtainable but with a more cumbersome algorithm. Nevertheless, herein we confine our attention to the convex case leaving the more general result to be reported upon elsewhere; however, see [10] wherein Jacobians in pieces of linearity are assumed to be nonsingular.

2. Notation and lemma

Let $\mathbf{R}^{m \times n}$ denote the set of real $m \times n$ matrices and let $\mathbf{R}^m = \mathbf{R}^{m \times 1}$. If $A \in \mathbf{R}^{m \times n}$, then by $A_i \in \mathbf{R}^{1 \times n}$ and $A_{.i} \in \mathbf{R}^m$ we denote the ith row and column of A, respectively. If $\alpha = \{i_1 < i_2 < \ldots < i_k\}$, then by $A_\alpha \in \mathbf{R}^{k \times n}$ and $A_{.\alpha} \in \mathbf{R}^{m \times k}$ we denote the submatrices $(A_{i_1}, \ldots, A_{i_k})$ and $(A_{.i_1}, \ldots, A_{.i_k})$ of rows and columns of A. By $A_{\alpha\beta}$ we denote $(A_\alpha)_{.\beta} \in \mathbf{R}^{k \times \ell}$, where $\beta = \{j_1 < \ldots < j_\ell\}$. If A is in $\mathbf{R}^m$, then by A_β we denote $A_{.\beta}$. Given a set α, we denote its size by $|\alpha|$. If $A \in \mathbf{R}^{m \times n}$ and $B \in \mathbf{R}^{n \times k}$, then by $A\,B \in \mathbf{R}^{m \times k}$ we denote the ordinary matrix product. By $A \geq 0$ and $A > 0$ we denote that each element of A is nonnegative and positive, respectively. By $A \neq 0$ we denote that some element of A is nonzero.

The following lemma, known as Farkas' Lemma, has an elementary proof via the Dines–Fourier–Motzkin elimination scheme using induction on the number of variables, see [4, 3, p. 84, and 17].

Lemma 2.1. *If* $\{x : A\,x \leq b\}$ *is empty, then there is a* $0 \leq \lambda \in \mathbf{R}^{1 \times m}$ *such that* $\lambda A = 0$, $\lambda b < 0$, *and* $\sum_1^m \lambda_i = 1$.

3. The form of F and F^ε

In this section we introduce the structures necessary to specify F as being the maximum of a finite collection of linear functions.

Let μ and v be the sets $\{1, \ldots, m\}$ and $\{1, \ldots, n\}$, where $m \geq n > 0$. Let $\mu_1, \ldots, \mu_n$ be a partition of μ. Let Q be a matrix in $\mathbf{R}^{m \times n}$ and r a vector in $\mathbf{R}^m$.

Define the real valued function $f_i : \mathbf{R}^n \to \mathbf{R}^1$ by

$$f_i(x) = \max_{j \in \mu_i} \{Q_j x + r_j\}$$

for each $i \in v$. Define $F : \mathbf{R}^n \to \mathbf{R}^n$ by $F(x) = (f_1(x), \ldots, f_n(x))$. Let $\Delta = \mathsf{X}_{i \in v}\, \mu_i$, then we also have

$$F(x) = \max_{\delta \in \Delta} \{Q_\delta x + r_\delta\}.$$

To assure convergence of the algorithm, a class of perturbations F^ε of F is utilized. The perturbation is based upon the initial regular point x_0 and some $\delta_0 \in \Delta$ with $F(x_0) = Q_{\delta_0} x_0 + r_{\delta_0}$. This perturbation scheme is essentially that used in the simplex method, see [3, p. 231].

Let $\beta_0 = \mu \sim \delta_0$ and let $[\varepsilon] \in \mathbf{R}^{m-n}$ be the vector of powers $(\varepsilon^1, \varepsilon^2, \ldots, \varepsilon^{m-n})$ of ε. Now define $r^\varepsilon \in \mathbf{R}^m$ by

$$r^\varepsilon_{\beta_0} = r_{\beta_0} - [\varepsilon], \qquad r^\varepsilon_{\delta_0} = r_{\delta_0}$$

and define $F^\varepsilon : \mathbf{R}^n \to \mathbf{R}^n$ for $\varepsilon \geq 0$ by

$$F^\varepsilon(x) = \max_{\delta \in \Delta} Q_\delta x + r^\varepsilon_\delta.$$

Clearly, $F^\varepsilon \leq F$, $F^\varepsilon(x_0) = Q_{\delta_0} x_0 + r_{\delta_0}$, and $F^0 = F$.

4. Subgradients of F

For a real-valued convex function $f : \mathbf{R}^n \to \mathbf{R}^1$, the vector $h \in \mathbf{R}^{1 \times n}$ is defined to be a subgradient of f at x if and only if

$$f(y) \geq f(x) + h(y - x)$$

for all $y \in \mathbf{R}^n$. Let $\partial f(x)$ denote the set of subgradients of f at x; see [24] for an extensive treatment of this notion. Hence $H \in \mathbf{R}^{n \times n}$ is a sub-gradient of F at x if and only if H_i is a subgradient of f_i at x for each $i \in v$. In other words, letting $\partial F(x)$ denote the set of subgradients of F at x, we have the familiar formula $\partial F(x) = \partial f_1(x) \times \ldots \times \partial f_n(x)$. This equality together with the following lemma instructs one how to obtain $\partial F(x)$ for a given x, see [12] for a more general statement of the lemma.

Lemma 4.1. *The vector h is a subgradient of f_i at $\bar{x}$ if and only if h is in the convex hull of $\{Q_j : j \in \mu_i(\bar{x})\}$, where*

$$\mu_i(\bar{x}) = \{j \in \mu_i : f_i(\bar{x}) = Q_j \bar{x} + r_j\}.$$

Proof. If $h \in \partial f_i(\bar{x})$, then

$$f_i(y) = \max_{j \in \mu_i} \{Q_j y + r_j\} \geq \max_{j \in \mu_i} \{Q_j \bar{x} + r_j + h(y - \bar{x})\}$$

for $y \in \mathbf{R}^n$. There is a neighborhood N of $\bar{x}$ such that for $y \in N$ we have

$$f_i(y) = \max_{j \in \mu_i(\bar{x})} \{Q_j y + r_j\} \geq \max_{j \in \mu_i(\bar{x})} \{Q_j \bar{x} + r_j + h(y - \bar{x})\}.$$

Hence we have

$$\max_{j \in \mu_i(\bar{x})} \{Q_j(y - \bar{x}) \geq h(y - \bar{x})\}$$

for $y \in N$. Therefore,

$$\max_{j \in \mu_i(\bar{x})} \{Q_j y \geq h y\}$$

for $y \in \mathbf{R}^n$. It follows that there is no solution of

$$(Q_j - h) y \leq -1 \qquad j \in \mu_i(\bar{x}).$$

Using Lemma 2.1, it follows that there is a solution $\lambda_j \geq 0$ to

$$\sum_{j \in \mu_i(\bar{x})} (Q_j - h) \lambda_j = 0 \quad \text{and} \quad \sum_{j \in \mu_i(\bar{x})} \lambda_j = 1.$$

Hence h is a convex combination of the Q_j with $j \in \mu_i(\bar{x})$. The "if" portion of the proof is obtained by "reversing" the argument just given.

Hence $H \in \partial F(\bar{x})$ if and only if for each $i \in v$, H_i is in the convex hull of $\{Q_j : j \in \mu(\bar{x})\}$.

5. An example of F

We pause here to exhibit the notation in an example; a function F is described which is regular. In Section 13 we show that Newton's algorithm cycles in an attempt to solve $F(x) = 0$ for this choice of F. With $n = 2$ and $m = 4$, let Q and r be the matrices

$$\begin{pmatrix} 1 & 1 \\ -2 & 1 \\ -1 & 1 \\ 2 & 1 \end{pmatrix}, \quad \begin{pmatrix} -2 \\ 1 \\ -2 \\ +1 \end{pmatrix}$$

and let $\mu_1 = \{1,2\}$ and $\mu_2 = \{3,4\}$. Hence

$$f_1(x) = \max \begin{cases} (1,1)x - 2, \\ (-2,1)x + 1, \end{cases} \qquad f_2(x) = \max \begin{cases} (-1,1)x - 2, \\ (2,1)x + 1, \end{cases}$$

and $F(x) = (f_1(x), f_2(x))$. It is easy to verify that

$$f_1(x) = \begin{cases} (1,1)\,x - 2 & \text{for } x_1 \geq 1, \\ (-2,1)\,x + 1 & \text{for } x_1 \leq 1, \end{cases}$$

$$f_2(x) = \begin{cases} (-1,1)\,x - 2 & \text{for } x_1 \leq -1, \\ (2,1)\,x + 1 & \text{for } x_1 \geq -1. \end{cases}$$

Using Lemma 4.1, we see that if $x_1 > 1$, then

$$\partial F(x) = \left\{ \begin{pmatrix} 1 & 1 \\ 2 & 1 \end{pmatrix} \right\};$$

if $x_1 = 1$, then

$$\partial F(x) = \left\{ \begin{pmatrix} 1 - 3\lambda & 1 \\ 2 & 1 \end{pmatrix} : \quad 0 \leq \lambda \leq 1 \right\};$$

if $-1 < x_1 < 1$, then

$$\partial F(x) = \left\{ \begin{pmatrix} -2 & 1 \\ 2 & 1 \end{pmatrix} \right\};$$

if $x_1 = -1$, then

$$\partial F(x) = \left\{ \begin{pmatrix} -2 & 1 \\ -1 + 3\lambda & 1 \end{pmatrix} : \quad 0 \leq \lambda \leq 1 \right\};$$

if $x_1 < -1$, then

$$\partial F(x) = \left\{ \begin{pmatrix} -2 & 1 \\ -1 & 1 \end{pmatrix} \right\}.$$

Observe that F is regular but that

$$\begin{pmatrix} 0 & 1 \\ 0 & 1 \end{pmatrix} \in \partial f_1(1,0) \times \partial f_2(-1,0)$$

is singular. The zeros of f_1 and f_2 are exhibited in Fig. 1; observe that F has a unique zero.

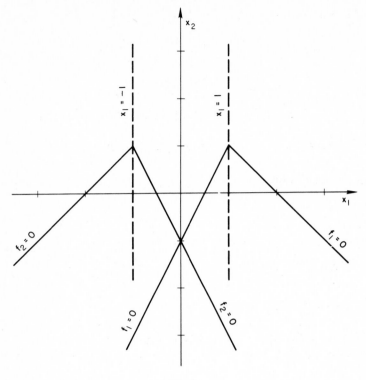

Fig. 1.

6. Local regularity and univalence under perturbation

We say that F^ε is univalent on N if $x \in N$, $y \in N$, and $F^\varepsilon(x) = F^\varepsilon(y)$ imply $x = y$. The lemma of this section shows that all $F^\varepsilon(x)$ for small $\varepsilon \geq 0$ are univalent on one neighborhood N of x_0 if F is regular at x_0.

Lemma 6.1. *If F is regular at x_0, then there is a neighborhood N of x_0 such that F^ε is univalent on N for $0 \leq \varepsilon \leq \varepsilon_0$ for some $\varepsilon_0 > 0$.*

Proof. If not, there is a sequence $(\varepsilon_i, x_i, y_i)$ with $\varepsilon_i \to 0$, $x_i \neq y_i$, $x_i \to x_0$, $y_i \to x_0$, and $F^{\varepsilon_i}(x_i) = F^{\varepsilon_i}(y_i)$ as $i \to +\infty$. It follows that there are δ_1 and δ_2 in Δ and an i with

$$Q_{\delta_1} x_0 + r_{\delta_1} = Q_{\delta_2} x_0 + r_{\delta_2} = F(x_0),$$

$$Q_{\delta_1} x_i + r_{\delta_1}^{\varepsilon_i} = Q_{\delta_2} y_i + r_{\delta_2}^{\varepsilon_i},$$

$$Q_{\delta_1} x_i + r_{\delta_1}^{\varepsilon_i} \geq Q_{\delta_2} x_i + r_{\delta_2}^{\varepsilon_i},$$

$$Q_{\delta_2} y_i + r_{\delta_2}^{\varepsilon_i} \geq Q_{\delta_1} y_i + r_{\delta_1}^{\varepsilon_i}.$$

By Lemma 4.1, Q_{δ_1} and Q_{δ_2} are subgradients of F at x_0 and

$$Q_{\delta_1}(x_i - y_i) \geq 0, \qquad Q_{\delta_2}(x_i - y_i) \leq 0.$$

Taking convex combinations of Q_{δ_1} and Q_{δ_2} row by row we get a Q with

$$Q(x_i - y_i) = 0$$

By Lemma 4.1, Q is a subgradient of F at x. But Q is regular and this is a contradiction.

7. The system $F^{\varepsilon}(x) = a + b\,z$

Our algorithm operates by generating a family of paths $(x^{\varepsilon}(\cdot),\ z^{\varepsilon}(\cdot))$ of solutions to the system

$$F^{\varepsilon}(x) = a + b\,z$$

for small $\varepsilon \geq 0$ (that is, small $\varepsilon > 0$ and $\varepsilon = 0$). In this section we specify the transform of this system used by the algorithm. Assuming F is regular at x_0 we set $a = F(x_0)$ and select $b \neq 0$ in $\mathbf{R}^n$ arbitrarily.

Define $a' = (a'_1, \ldots, a'_m) \in \mathbf{R}^m$ and $b' = (b'_1, \ldots, b'_m) \in \mathbf{R}^m$ by

$$a'_j = a_i, \qquad b'_j = b_i \quad \text{for } j \in \mu_i,$$

where $a = (a_1, \ldots, a_n)$ and $b = (b_1, \ldots, b_n)$. Let $I \in \mathbf{R}^{m \times m}$ be the identity and consider the system $(1,\varepsilon)$:

a) $\qquad I\,w + Q\,x - b'z = a' - r^{\varepsilon},$

b) $\qquad 0 \not< w_{\mu_i} \geq 0 \quad \text{for all } i \in v,$ $\qquad\qquad (1,\varepsilon)$

where $w = (w_1, \ldots, w_m) \in \mathbf{R}^m$ and where $0 \not< w_{\mu_i} \geq 0$ denotes that $w_j \geq 0$ for all $j \in \mu_i$ and that $w_j = 0$ for some $j \in \mu_i$.

Lemma 7.1. *If (x, z) solves $F^\varepsilon(x) = a + b\,z$ and w is obtained from $(1, \varepsilon, a)$, then (w, x, z) solves $(1, \varepsilon)$; if, in addition, $\delta \in \Delta$ and $F^\varepsilon(x) = Q_\delta x + r^\varepsilon$, then $w_\delta = 0$. If (w, x, z) solves $(1, \varepsilon)$, then (x, z) solves $F^\varepsilon(x) = a + b\,z$; if, in addition, $\delta \in \Delta$ and $w_\delta = 0$, then $F^\varepsilon(x) = Q_\delta x + r^\varepsilon_\delta$.*

Proof. The conclusion follows from the fact that

$$\max_{j \in \mu_i} \{Q_j x + r^\varepsilon_j\} - b_i z = a_i$$

if and only if $0 \not< w_{\mu_i} \geq 0$, where

$$w_j = a_i + b_i z - Q_j x - r^\varepsilon_j$$

for $j \in \mu_i$.

We could describe the algorithm as applied to $(1, \varepsilon)$ as in [8]; however, it is perhaps computationally and conceptually simpler to eliminate the x variables.

Recalling $\beta_0 = \mu \sim \delta_0$, rewrite $(1, \varepsilon)$ in two parts, $(2, \varepsilon)$ and (3).

a) $\qquad I_{\beta_0} w + Q_{\beta_0} x - b'_{\beta_0} z = a'_{\beta_0} - r^\varepsilon_{\beta_0},$

b) $\qquad 0 \not< w_{\mu_i} \geq 0 \quad$ for $i \in v,$ $\qquad\qquad\qquad (2, \varepsilon)$

$$x = x_0 + Q_{\delta_0}^{-1}(b\,z - w_{\delta_0}). \qquad\qquad\qquad (3)$$

Hence we can use (3) to eliminate x from $(2, \varepsilon)$ to obtain

a) $\qquad M w + d\,z = c + [\varepsilon],$

b) $\qquad 0 \not< w_{\mu_i} \geq 0 \quad$ for $i \in v,$ $\qquad\qquad\qquad (4, \varepsilon)$

where

$$M = I_{\beta_0} - Q_{\beta_0} Q_{\delta_0}^{-1} I_{\delta_0},$$

$$d = -b'_{\beta_0} + Q_{\beta_0} Q_{\delta_0}^{-1} b,$$

$$c = a'_{\beta_0} - r_{\beta_0} - Q_{\beta_0} x_0.$$

The system $(4, \varepsilon)$ is essentially that used by Cottle and Dantzig [2]; Lemke [19] has shown that it can be put into the form of the linear complementary problem. As with $(1, \varepsilon)$ we state the importance of (3) and $(4, \varepsilon)$ in a lemma.

Lemma 7.2. *If (w, z) solves $(4, \varepsilon)$ and x is obtained from (3), then (x, z) solves $F^\varepsilon(x) = a + b\, z$; if, in addition, $\delta \in \Delta$ and $w_\delta = 0$, then $F^\varepsilon(x) = Q_\delta y + r_\delta^\varepsilon$. If (x, z) solves $F^\varepsilon(x) = a + b\, z$ and w is obtained from $(1, \varepsilon, a)$, then (w, z) solves $(4, \varepsilon)$; if, in addition, $\delta \in \Delta$ and $F^\varepsilon(x) = Q_\delta x + r_\delta^\varepsilon$, then $w_\delta = 0$.*

Observe, using the lemma, that if we have a path of solutions $(w^\varepsilon(\cdot), z^\varepsilon(\cdot))$ to $(4, \varepsilon)$, then upon using (3) we obtain a path of solutions $(x^\varepsilon(\cdot), z^\varepsilon(\cdot))$ to $F^\varepsilon(x) = a + b\, z$. Our algorithm generates a path of solutions to $(4, \varepsilon)$ for small $\varepsilon \geq 0$.

Lemma 7.3. *Given $(4, \varepsilon)$ we have $M_{\cdot \beta_0} = I \in \mathbf{R}^{(m-n) \times (m-n)}$ and $c \geq 0$.*

Proof. Since the columns of I_{δ_0} indexed by $\beta_0 = \mu \sim \delta_0$ are zero, the first conclusion follows from the definition of M. Using $(x_0, 0)$ and $F^\varepsilon(x_0) = Q_{\delta_0}(x_0) + r_{\delta_0} = a$ and the second part of Lemma 7.2, we get $w \geq 0$. From $(4, \varepsilon)$ we get $c = w_{\beta_0} \geq 0$.

Of course, the existence of $(4, \varepsilon)$ is based upon the assumption that x_0 is regular.

8. Bases of $M\, w + d\, z = c + [\varepsilon]$

The algorithm generates a path of solutions $(w^\varepsilon(\cdot), z^\varepsilon(\cdot))$ to

a) $\qquad M\, w + d\, z = c + [\varepsilon],$

b) $\qquad 0 \not< w_{\mu_i} \geq 0 \quad$ for $i \in v$ $\qquad\qquad (4, \varepsilon)$

by pivoting from basis to basis. Those readers familiar with complementarity pivot theory will find this section quite routine, and consequently, proofs are usually omitted.

Definition 8.1. A subset β of $\mu \cup \{m + 1\}$ is defined to be a basis[1] if

(i) $|\beta| = m - n$,
(ii) $\beta \cap \mu_i \neq \mu_i \quad$ for $i \in v$,
(iii) $(M, d)_{\cdot \beta}^{-1}$ exists,
(iv) $(M, d)_{\cdot \beta}^{-1}(c + [\varepsilon]) \geq 0$ for small $\varepsilon > 0$.

---

[1] That is, a nondegenerate basis index.

If β is a basis and

$$(w, z)_\beta = (M, d)^{-1}_{\cdot\,\beta}(c + [\varepsilon]) \geq 0,$$
$$(w, z)_{\sim\beta} = 0,$$

then (w, z) is defined to be an ε-basic solution. Clearly, an ε-basic solution (w, z) solves $(4, \varepsilon)$. From Lemma 7.3, we see that $\beta_0 = \mu \sim \delta_0$ is a basis; indeed, it is this basis with which we initiate the algorithm.

Lemma 8.2. *There is an $\varepsilon_0 > 0$ such that for any $0 < \varepsilon \leq \varepsilon_0$ and basis β we have*

$$(M, d)^{-1}_{\cdot\,\beta}(c + [\varepsilon]) > 0.$$

Proof. See [3, p. 232].

If β is a basis, then either (a) or (b).
(a) $m + 1 \notin \beta$. In this case, $|\beta \cap \mu_i| = |\mu_i| - 1$ for all $i \in v$.
(b) $m + 1 \in \beta$. In this case, for exactly one $j \in v$, $|\beta \cap \mu_j| = |\mu_j| - 2$ and for $i \in v \sim \{j\}$, $|\beta \cap \mu_i| = |\mu_i| - 1$.
We define $\bar\beta = \{m + 1\}$ for case (a) and $\bar\beta = \mu_j \sim \beta$ in case (b). Note in case (b) that $|\bar\beta| = 2$.

Definition 8.3. Two distinct bases α and β are defined to be adjacent if

$$\beta = \alpha \cup \{k\} \sim \{\ell\}$$

for some $k \in \bar\alpha$ and $\ell \in \alpha$ or equivalently, if

$$\alpha = \beta \cup \{k\} \sim \{\ell\}$$

for some $k \in \bar\beta$ and $\ell \in \beta$.

Lemma 8.4. *If α and β are adjacent bases and if (w, z) and (w', z') are corresponding ε-basic solutions, then*

$$(1 - \lambda)(w, z) + \lambda(w', z')$$

for each $0 \leq \lambda \leq 1$ is a solution of $(4, \varepsilon)$.

Lemma 8.5. *If β is a basis and $i \in \bar{\beta}$, then there is at most one basis of form*

$$\beta \cup \{i\} \sim \{j\},$$

where $j \in \beta$. Hence β is adjacent to at most $|\bar{\beta}|$ bases.

Lemma 8.6. *If β is a basis, if $i \in \bar{\beta}$, and there is no basis of form $\beta \cup \{i\} \sim \{j\}$, where $j \in \beta$, then*

$$(w, z) + s\,(w', z')$$

for every $s \geq 0$ is a solution of $(4, \varepsilon)$, where (w, z) is the ε-basic solution of β and where

$$\begin{pmatrix} w' \\ z' \end{pmatrix}_i = 1, \qquad \begin{pmatrix} w' \\ z' \end{pmatrix}_\beta = -(M, d)_{\cdot\beta}^{-1} M_i, \qquad \begin{pmatrix} w' \\ z' \end{pmatrix}_k = 0$$

for $k \notin \beta \cup \{i\}$.

9. The algorithm

Here we specify an algorithm for generating paths of solutions (w, z) with $z \geq 0$ to the system

a) $\qquad M w + d z = c + [\varepsilon],$

b) $\qquad 0 \nleq w_{\mu_i} \geq 0 \qquad i \in v.$ $\hfill (4, \varepsilon)$

The algorithm begins with the bases β_0 and iterates to new adjacent bases as long as possible.

Algorithm
  (1) Let $\beta_{-1} = \beta_0$ and $k = 0$.
  (2) Assume that the sequence of bases $\beta_0, \ldots, \beta_k$ has been generated. Determine if there is a basis β of form

$$\beta_k \cup \{i\} \sim \{j\},$$

where $i \in \bar{\beta}_k \sim \beta_{k-1}$ and $j \in \beta_k$. If so, let $\beta_{k+1} = \beta$, increase k by 1, and go to (2); if not, terminate.

In the statement of the algorithm, we could equivalently have said "Determine if there is a basis $\beta \neq \beta_{k-1}$ which is adjacent to $\beta_k, \ldots$". It is important to know that the operations of seeking β_{k+1} given β_k is computationally simple and is essentially that of a feasible pivot as in linear programming.

The theorem describes the output of the algorithm. It will also be routine to those familiar with complementary pivot theory. This type of theorem first appeared in [20 and 18].

Theorem 9.1. *The sequence of basic indices $\beta_0, \ldots, \beta_\ell$ generated by the algorithm is finite, unique, and its elements are distinct. Either*

(I) $\bar{\beta}_\ell \sim \beta_{\ell-1} = \emptyset$ *and* $\bar{\beta}_\ell = \{m+1\}$,
(II) $\bar{\beta}_\ell \sim \beta_{\ell-1} = \{i\}$ *and there is no basis of form* $\beta_\ell \cup \{i\} \sim \{j\}$ *with* $j \in \beta_\ell$.

Proof. The proof follows from Lemma 8.5 and the facts that adjacency is symmetric, that $|\bar{\beta}_0| = 1$ and $|\bar{\beta}_k| \leq 2$, and that there are only finitely many bases.

We shall speak of the algorithm as terminating without and with a ray corresponding to the termination conditions (I) and (II), see Lemma 8.6. We use ℓ, henceforth, to indicate the number of iterations the algorithm required to terminate.

10. The generated paths $(x^\varepsilon(\cdot), z^\varepsilon(\cdot))$

Let $\beta_0, \ldots, \beta_\ell$ be the sequence of bases generated by the algorithm. We show that this sequence yields a family (indexed by ε) of paths $(x^\varepsilon(\cdot), z^\varepsilon(\cdot))$ of solutions to $F^\varepsilon(x) = a + bz$ for small $\varepsilon \geq 0$.

Select $\varepsilon_0 > 0$ according to Lemma 8.2 and let $(w^\varepsilon(k), z^\varepsilon(k))$ be the ε-basic solution corresponding to β_k for $k = 0, 1, \ldots, \ell$ and $0 \leq \varepsilon \leq \varepsilon_0$. First extend the domain of $(w^\varepsilon(\cdot), z^\varepsilon(\cdot))$ from $\{0, 1, \ldots, \ell\}$ to $[0, \ell]$ by setting $(w^\varepsilon(t), z^\varepsilon(t))$ to be

$$(k + 1 - t)(w^\varepsilon(k), z^\varepsilon(k)) + (t - k)(w^\varepsilon(k+1), z^\varepsilon(k+1))$$

for $k \leq t \leq k+1$ for $k = 0, \ldots, \ell - 1$. If the algorithm terminated with a ray, then extend the domain of $(w^\varepsilon(\cdot), z^\varepsilon(\cdot))$ from $[0, \ell]$ to $[0, +\infty)$ by setting $(w^\varepsilon(t), z^\varepsilon(t))$ equal to

$$(w^\varepsilon(\ell), z^\varepsilon(\ell)) + (t - \ell)(w', z')$$

for $t \geq \ell$, where (w', z') is described in Lemma 8.6, where $\beta = \beta_\ell$ and $\{i\} = \bar{\beta}_\ell \sim \beta_{\ell-1}$. Let $J = [0, \ell]$ or $J = [0, +\infty)$ corresponding to termination without and with a ray. Finally, $(w^\varepsilon(t), z^\varepsilon(t))$ with $t \in J$ is a path of solutions to $(4, \varepsilon)$ for $0 \leq \varepsilon \leq \varepsilon_0$.

Now use $(w^\varepsilon(\cdot), z^\varepsilon(\cdot))$, (3), and Lemma 7.2 to get the path $(x^\varepsilon(\cdot), z^\varepsilon(\cdot))$ for each $0 \leq \varepsilon \leq \varepsilon_0$. We have

$$F^\varepsilon(x^\varepsilon(t)) = a + b \, z^\varepsilon(t)$$

for $t \in J$ and $0 \leq \varepsilon \leq \varepsilon_0$. We shall refer to these paths $(x^\varepsilon(\cdot), z^\varepsilon(\cdot))$ as those generated by the algorithm. The next lemmas record basic facts about these paths; the first describes the starting point of the path, the second discusses the generated path as related to terminal conditions of the algorithm, the third demonstrates that the path contains no loops, the fourth shows that $x^0(\ell) \neq x_0$ if the algorithm terminates without a ray, and the fifth relates the δ's to the path.

Lemma 10.1. $(x^\varepsilon(0), z^\varepsilon(0)) = (x_0, 0)$ *for all* $0 \leq \varepsilon \leq \varepsilon_0$. $z^\varepsilon(1) > 0$ *for all* $0 < \varepsilon \leq \varepsilon_0$.

Lemma 10.2. *If the algorithm terminates without a ray, then* $z^\varepsilon(\ell) = 0$ *for* $0 \leq \varepsilon \leq \varepsilon$. *If the algorithm terminates with a ray then*

$$(x^\varepsilon(t), z^\varepsilon(t)) = (x^\varepsilon(\ell), z^\varepsilon(\ell)) + (t - \ell)(x', z'),$$

where $(x', z') \neq 0$ *for* $0 \leq \varepsilon \leq \varepsilon_0$ *and* $t \geq \ell$.

Proof. If $m + 1 \notin \beta_\ell$, then $z^\varepsilon(\ell) = 0$ in view of the definition of an ε-basic solution. From Lemma 8.6, we have $(w', z') \neq 0$. From $(1, \varepsilon)$, we have

$$I w' + Q x' - b z' = 0.$$

If $z' = 0$, then $w' \neq 0$, and we have $x' \neq 0$.

Lemma 10.3. *For each $0 < \varepsilon \leq \varepsilon_0$ the path $(x^\varepsilon(\cdot), z^\varepsilon(\cdot))$ is univalent on J.*

Proof. If $(x^\varepsilon(t), z^\varepsilon(t))$ equals $(x^\varepsilon(s), z^\varepsilon(s))$, then $(w^\varepsilon(t), z^\varepsilon(t))$ equals $(w^\varepsilon(s), z^\varepsilon(s))$. Let $\gamma \subset \mu \cup \{m + 1\}$ index the positive components of $(w^\varepsilon(t), z^\varepsilon(t))$. From Lemma 8.2, γ contains at least one basis β and from Lemmas 8.2 and 8.5, γ contains at most two bases. Note that $\gamma = \beta_k$ or that $\gamma = \beta_k \cup \{i\}$, where $\{i\} = \bar{\beta}_k \sim \beta_{k-1}$. The result now follows from the definition of ε-basic solution and the derived paths.

Lemma 10.4. *If the algorithm terminates without a ray, then $x^0(\ell) \neq x_0$.*

Proof. By Lemma 10.1, we have

$$F^\varepsilon(x^\varepsilon(\ell)) = a$$

for $0 \leq \varepsilon \leq \varepsilon_0$. According to Lemma 6.1, there is a neighborhood N of x_0 on which F^ε is univalent for $0 \leq \varepsilon \leq \varepsilon_1$ for some $\varepsilon_1 > 0$. In view of Lemma 10.2, we have $x^\varepsilon(\ell) \neq x_0$ for small $\varepsilon > 0$. Hence $x^\varepsilon(\ell) \notin N$ for small $\varepsilon > 0$. Therefore, $x^0(\ell) \neq x_0$.

Lemma 10.5. *For $k = 0, \ldots, \ell$ let $\delta_k = \mu \sim (\beta_k \cup \{i\}) \in \Delta$, where $i \in \bar{\beta}_k \sim \beta_{k-1}$. For $0 \leq \varepsilon \leq \varepsilon^0$ the expression*

$$F^\varepsilon(x^\varepsilon(t)) = Q_{\delta_k} x^\varepsilon(t) + r^\varepsilon_{\delta_k}$$

is obtained for $k = 0, \ldots, \ell - 1$ and $k \leq t \leq k + 1$, for $t = k = \ell$, and if the algorithm terminates with a ray for $t \geq k = \ell$.

11. An example of the algorithm

Let F be the regular function described in Section 5. We use the algorithm to solve $F(x) = a + bz$ for all $z \geq 0$; as we shall show in Section 13, such results are assurred when F is regular. Throughout the example we assume that the reader is familiar with the notion of a pivot as used in the simplex method, see [3, p. 109].

Let $x(0) = (2,1)$, $a = F(2,1) = (1,6)$, and $\delta_0 = \{1,4\}$. Assuming that we would like to compute the zero of F we set $b = -a$; a solution $(x,1)$ of $F(x) = a + bz$ yields x a zero.

The system $(1, \varepsilon, a)$ in detached coefficient form can be represented as

w_1	w_2	w_3	w_4	x_1	x_2	z		
1				1	1	1		3
	1			-2	1	1		0
		1		-1	1	6	$=$	8
			1	2	1	6		5

Block pivot on rows $\{1,4\}$ and columns $\{5,6\}$ to obtain

w_1	w_2	w_3	w_4	x_1	x_2	z		
-1	0	0	1	1	0	5		2
-4	1	0	3	0	0	15		3
-3	0	1	2	0	0	15	$=$	9
2	0	0	-1	0	1	-4		1

Hence (3) becomes

		z	w_1	w_4	
x_1	$=$	2	-5	1	-1
x_2		1	4	-2	$+1$

and $(4, \varepsilon, a)$ becomes

w_1	w_2	w_3	w_4	z		
-4	1	0	3	15	$=$	3
-3	0	1	2	15		9

both in detached coefficient form. This last system can appropriately be described as the β_0-tableau. From this tableau one can read off the corresponding ε-basic solution, in particular, $(w^\varepsilon(0), z^\varepsilon(0)) = (0, 3 + \varepsilon, 9 + \varepsilon^2, 0, 0)$. The coefficients of the ε terms are found in the corresponding rows of the columns listed in β_0.

Since $\bar{\beta}_0 = \{5\}$ and $3/15 < 9/15$ we have $\beta_1 = \{3,5\}$. Pivoting on column 1 and row 5 of the β_0-tableau, we get the β_1-tableau:

w_1	w_2	w_3	w_4	z		
$-4/15$	$1/15$	0	$3/15$	1	$=$	$3/15$
1	-1	1	-1	0		6

The ε-basic solution of β_1 is

$$(w^\varepsilon(1), z^\varepsilon(1)) = (0, 0, 6 - \varepsilon + \varepsilon^2, 0, 3/15 + \varepsilon/15).$$

Since $\overline{\beta}_1 \sim \beta_0 = \{1,2\} \sim \{2,3\} = \{1\}$, $1 > 0$, and $-4/15 < 0$, we have $\beta_2 = \{1,5\}$. Pivoting on column 1 and row 2 of the β_1-tableau we get the β_2-tableau:

w_1	w_2	w_3	w_4	z		
0	$-3/15$	$4/15$	$-1/15$	1	$=$	$27/15$
1	-1	1	-1	0		6

The ε-basic solution is

$$(w^\varepsilon(2), z^\varepsilon(2)) = (6 - \varepsilon + \varepsilon^2, 0, 0, 0, 27/15 - \varepsilon 3/15 + \varepsilon^2 4/15).$$

Since $\overline{\beta}_2 \sim \beta_1 = \{3,4\} \sim \{3,5\} = \{4\}$ and column 4 of the β_2-tableau is nonpositive, the algorithm terminates with a ray. We get $(w', z') = (1, 0, 0, 1, 1/15)$.

The path $(x^0(t), z^0(t))$ for $t \in J = [0, +\infty]$ is

$$(2 - t, 1 + 4t/5, t/5) \quad \text{for } 0 \leq t \leq 1,$$

$$(3 - 2t, 37/5 - 28t/5, -21/15 + 24t/15) \quad \text{for } 1 \leq t \leq 2,$$

$$(-1/3 - t/3, -35/15 - 11t/15, 25/15 + t/15) \quad \text{for } 2 \leq t.$$

In particular, if our purpose was to compute a zero of F, we observe that $z^0(3/2) = 1$ hence $x^0(3/2) = (0, -1)$ is a zero. The generated path is exhibited in Fig. 2.

In general, the task of moving from one β-tableau to another can be described as (i) running the "minimum ratio test" first with respect to the last column, and subsequently with respect to the columns listed in β_0 to resolve ties and (ii) pivoting on the indicated element.

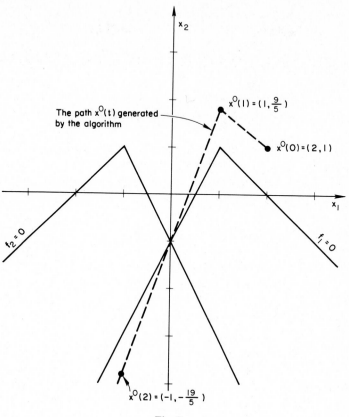

The path $x^0(t)$ generated by the algorithm

$x^0(1) = (1, \frac{9}{5})$

$x^0(0) = (2, 1)$

$f_2 = 0$

$f_1 = 0$

$x^0(2) = (-1, -\frac{19}{5})$

Fig. 2.

12. The main theorem

We now summarize the performance of the algorithm in the following theorem. We assume that F is regular at x_0, that $a = F(x_0)$, and $b \neq 0$.

Theorem 12.1 *The path $(x^0(\cdot), z^0(\cdot))$ generated by the algorithm begins at $(x_0, 0)$ and contains at least one of:*
  (a) *A solution $x \neq x_0$ to $F(x) = a$.*
  (b) *A ray of solutions to $F(x) = a + b\,z$ for some $z \geq 0$.*
  (c) *A solution $F(x) = a + b\,z$ for every $z \geq 0$.*

Proof. If the algorithm terminates without a ray, then $z^0(\ell) = 0$ and $x^0(\ell) \neq x_0$ according to Lemmas 10.2 and 10.4. If the algorithm terminates with a ray, then according to Lemma 10.2 we have either (b) or (c).

Of course, condition (c) occurs if and only if $z^0(t) \to +\infty$ as $t \to +\infty$.

Corollary 12.2. *If $z^\varepsilon(t)$ does not strictly increase in t for each sufficiently small $\varepsilon > 0$, then the path $x^0(\cdot)$ contains a point at which F is not regular.*

Proof. If $z^\varepsilon(t)$ does not strictly increase in t for small $\varepsilon > 0$, there is a $\bar{t}$ such that $z^\varepsilon(\cdot)$ is not univalent in any neighborhood of $\bar{t}$ for all small $\varepsilon > 0$. Hence in any neighborhood of $x^\varepsilon(\bar{t})$ for small $\varepsilon > 0$ there are $x^\varepsilon(t) \neq x^\varepsilon(s)$ with

$$F^\varepsilon(x^\varepsilon(t)) = F^\varepsilon(x^\varepsilon(s)).$$

From Lemma 6.1 it follows that F is not regular at $x^0(\bar{t})$.

The $\bar{t}$ of the proof is easily computed given the paths $z^\varepsilon(\cdot)$. Also observe that if $z^\varepsilon(t)$ strictly increases in t, then $z^\varepsilon(t) \to +\infty$ as $t \to +\infty$ for $\varepsilon > 0$; the latter implies that $z^0(t) \to +\infty$ as $t \to +\infty$.

13. Regular F

For regular F, we see that the path generated by the algorithm contain ε solution $x^0(\bar{t})$ to $F(x) = y$ providing we set $b = y - a$; from Corollary 12.2 we see that $z^0(t) \to +\infty$ hence $z^0(\bar{t}) = 1$ for some $\bar{t}$.

Since $z^\varepsilon(\cdot)$ increase for regular F, it is interesting to note that $(4,\varepsilon)$ can be solved for $z^0(t) \to +\infty$ by the simplex method applied to

$$\begin{aligned} \text{maximize} \quad & z, \\ \text{subject to} \quad & M\,w + d\,z = c, \\ & w \geq 0, \qquad z \geq 0. \end{aligned}$$

The starting basis would be β_0 and the basis entry choice among the columns with a positive relative cost coefficient must preserve $w_{\mu_i} \not> 0$ for $i \in v$.

For our next result we drop the assumption that F is piecewise linear; we retain the assumption that F is convex and that its subgradients are nonsingular. By F convex we mean that

$$F((1 - \lambda) x + \lambda y) \leq (1 - \lambda) F(x) + \lambda F(y)$$

for all x, y, and $0 \leq \lambda \leq 1$; subgradients and regularity are defined as before. If $\| G(x) \| \to +\infty$ as $\| x \| \to +\infty$, we write $G(\infty) = \infty$.

Our method of proof follows that of Fujisawa and Kuh [9, 10], namely we use the following immediate corollary of Palais' theorem [23, p. 128], also see [1, p. 156 and 14, p. 544].

Corollary 13.1 *If $G : \mathbf{R}^n \to \mathbf{R}^n$ is continuous, locally univalent, and $G(\infty) = \infty$, then G is a homeomorphism.*

Theorem 13.2. *If $F : \mathbf{R}^n \to \mathbf{R}^n$ is regular and $F(\infty) = \infty$, then F is a homeomorphism.*

Proof. In view of [24, p. 82] and the corollary, we need only show that F is locally univalent. If $x_i \neq y_i$ tend to x and

$$F(x_i) = F(y_i),$$

then

$$F(x_i) \geq F(y_i) + Q_i(x_i - y_i),$$
$$F(y_i) \geq F(x_i) + P_i(y_i - x_i),$$

where Q_i and P_i are subgradients of F at y_i and x_i, see [24, p. 217]. Hence

$$P_i z_i \geq 0 \geq Q_i z_i,$$

where $z_i = x_i - y_i \neq 0$. As in the proof of Lemma 6.1, we take convex combinations of P_i and Q_i row by row to get $R_i = (I - \Lambda_i) P_i + \Lambda_i Q_i$, where $R_i z_i = 0$, $\det R_i = 0$, and $0 \leq \Lambda_i \leq I$. Via [24, pp. 234, 237] there are subgradients P and Q of F at x so that $P_i \to P$, $Q_i \to Q$, and $\Lambda_i \to \Lambda$ on a subsequence. Then

$$R_i \to (I - \Lambda) P + \Lambda Q = R$$

and det $R = 0$. By [24, p. 215], R is a subgradient of F at x and F is not regular at x.

It would be interesting to pursue the relation between this theorem and those of Fujisawa and Kuh [9, p. 502] and Nikaido [21, p. 355].

If the determinants of all subgradients of convex F are bounded away from zero, we say that F is totally regular.

Conjecture. If F is totally regular, then F is a homeomorphism.

One might believe that if F is regular on a convex set S, then F is univalent on that set; we show that this is not so.

With $n = 2$ and $m = 4$ let Q and r be the matrices

$$\begin{pmatrix} -1 & 1 \\ 3 & 6 \\ 1 & 1 \\ -3 & 6 \end{pmatrix}, \quad \begin{pmatrix} -2 \\ -6 \\ -2 \\ -6 \end{pmatrix}$$

and let $\mu_1 = \{1,2\}$ and $\mu_2 = \{3,4\}$. Let $F = (f_1, f_2)$, where

$$f_1(x) = \max \{(-1, 1) x - 2, (3, 6) x - 6\},$$

$$f_2(x) = \max \{(1, 1) x - 2, (-3, 6) x - 6\}.$$

Using Lemma 4.1 and examining $\partial F(x)$ on the set

$$P = \{x: -2 \le x_1 \le 2, x_2 = 0\},$$

we see that if $-2 \le x_1 < -1$, then

$$\partial F(x) = \left\{ \begin{pmatrix} -1 & 1 \\ -3 & 6 \end{pmatrix} \right\};$$

if $x_1 = -1$, then

$$\partial F(x) = \left\{ \begin{pmatrix} -1 & 1 \\ 1 - 4\lambda & 1 + 5\lambda \end{pmatrix} : 0 \le \lambda \le 1 \right\};$$

if $-1 < x_1 < 1$, then

$$\partial F(x) = \left\{ \begin{pmatrix} -1 & 1 \\ 1 & 1 \end{pmatrix} \right\};$$

if $x_1 = 1$, then

$$\partial F(x) = \left\{ \begin{pmatrix} -1 + 4\lambda & 1 + 5\lambda \\ 1 & 1 \end{pmatrix} : 0 \le \lambda \le 1 \right\};$$

if $1 < x_1 \le 2$, then

$$\partial F(x) = \left\{ \begin{pmatrix} 3 & 6 \\ 1 & 1 \end{pmatrix} \right\}.$$

Hence F is regular on P, nevertheless, $F(-2, 0) = 0 = F(2, 0)$. In Fig. 3 we exhibit the zeros of F.

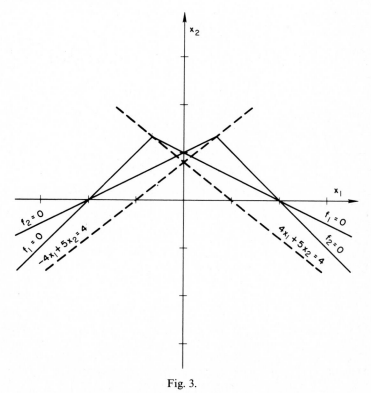

Fig. 3.

One might attempt to solve $F(x) = 0$ for a regular F by applying Newtons method. After describing Newtons method, we show that it cycles when applied to the example of Section 5.

Newtons Algorithm

    (1) Select $x(0) \in \mathbf{R}^n$ and set $i = 0$.
    (2) Select $\delta_i \in \Delta$ with $Q_{\delta_i} x(i) + r_{\delta_i} = F(x(i))$.
    (3) Solve $Q_{\delta_i} x(i + 1) + r_{\delta_i} = 0$.
    (4) Increase i by 1 and go to (2).

Letting $x(0) = (2,1)$ for the example of Section 5, we get $x(1) = (-3, 5)$, $x(2) = (3, 5)$, $x(3) = (-3, 5), \ldots$ and the estimate forever alternates between $x(1)$ and $x(2)$.

Sufficient conditions assuring that Newtons method converges globally for a regular function F are, essentially, that H^{-1} exist and be nonnegative for all $H \in \partial F(\mathbf{R}^n)$, see [22, p. 453].

References

[1] E. Artin and H. Braun, *Introduction to algebraic topology* (Charles E. Merrill, Columbus, Oh., 1969).
[2] R.W. Cottle and Dantzig, G.B., "A generalization of the linear complementarity problem", *Journal of Combinatorial Theory* 8 (1) (1970).
[3] G.B. Dantzig, *Linear programming and extensions* (Princeton University Press, Princeton, N.J., 1963).
[4] L.L. Dines, "Systems of linear inequalities", *Annals of Mathematics*, 2nd Ser. 20 (1919).
[5] B.C. Eaves, "An odd theorem", *Proceedings of the American Mathematical Society*, 26 (3) (1970) 509–513.
[6] B.C. Eaves, "Computing Kakutani fixed points", *SIAM Journal on Applied Mathematics* 21 (2) (1971) 236–244.
[7] B.C. Eaves, "On the basic theorem of complementarity", *Mathematical Programming* 1 (1) (1971) 68–75.
[8] B.C. Eaves, "A fixed point theorem from dynamic programming", Tech. Rept., Department of Operations Research, Stanford University, Stanford, Calif. (January 1973).
[9] T. Fujisawa and E.S. Kuh, "Some results on existence and uniqueness of solutions of nonlinear networks", *IEEE Transactions on Circuit Theory* 18 (5) (1971) 501–506.
[10] T. Fujisawa and E.S. Kuh, "Piecewise-linear theory of nonlinear networks", *SIAM Journal on Applied Mathematics* 22 (2) (1972) 307–328.
[11] T. Fujisawa, E.S. Kuh and T. Ohtsuki, "A sparse matrix method for analysis of piecewise-linear resistive networks", *IEEE Transactions on Circuit Theory* 19 (6) (1972) 571–584.

[12] R.C. Grinold, "Lagrangian subgradients", *Management Science* 17 (3) (1970) 185–188.

[13] M.W. Hirsch, "A proof of the nonretractibility of a cell onto its boundary", *Proceedings of the American Mathematical Society* 14 (1963) 364–365.

[14] C. Holzmann, and R.W. Liu, "On the dynamical equations of nonlinear networks with n-coupled elements", in: *Proceedings of the third annual allerton conference on circuit and system theory*, October 20–22, 1965, University of Illinois, Urbana, Ill., pp. 536–545.

[15] J. Katzenelson, "An algorithm for solving nonlinear resistor networks", *Bell Telephone Technical Journal* 44 (1965) 1605–1620.

[16] E.S. Kuh and I.N. Hajj, "Nonlinear circuit theory: Resistive networks", *Proceedings of the IEEE* 59 (3) (1971) 340–355.

[17] H.W. Kuhn, "Solvability and consistency for linear equations and inequalities", *American Mathematical Monthly* LXIII (4) (1956).

[18] C.E. Lemke, "Bimatrix equilibrium points and mathematical programming", *Management Science* 11 (1965) 681–689.

[19] C.E. Lemke, "Recent results on complementarity problems", in: *Nonlinear programming*, Eds. J.B. Rosen, O.L. Mangasarian and K. Ritter (Academic Press, New York, 1970).

[20] C.E. Lemke and J.T. Howson, Jr., "Equilibrium points of bimatrix games", *SIAM Journal on Applied Mathematics* 12 (1964) 413–423.

[21] H. Nikaido, *Convex structures and economic theory* (Academic Press, New York, 1968).

[22] J.M. Ortega and W.C. Rheinboldt, *Iterative solution of nonlinear equations in several variables* (Academic Press, New York, 1970).

[23] R.S. Palais, "Natural operations on differential forms", *Transactions of the American Mathematical Society* 92 (1959).

[24] R.T. Rockafellar, *Convex analysis* (Princeton University Press, Princeton, N.J., 1970).

[25] H.E. Scarf, "An algorithm for a class of nonconvex programming problems", Cowles Foundation Discussion Paper No. 211, Yale University, New Haven, Conn., 1966.

Mathematical Programming Study 1 (1974) 120–132 North-Holland Publishing Company

ON BALANCED MATRICES

D.R. FULKERSON*

Cornell University, Ithaca, N.Y., U.S.A.

A.J. HOFFMAN**

IBM Thomas J. Watson Research Center, Yorktown Heights, N.Y., U.S.A.

and

Rosa OPPENHEIM

Rutgers University, New Brunswick, N.J., U.S.A.

Received 14 February 1974
Revised manuscript received 15 April 1974

Dedicated to A.W. Tucker, as a token of our gratitude for
over 40 years of friendship and inspiration

1. Introduction

In his interesting paper [2], Claude Berge directs our attention to two questions relevant to the use of linear programming in combinational problems. Let A be a $(0, 1)$-matrix, w and c nonnegative integral vectors, and define the polyhedra

$$P(A, w, c) = \{y : y A \geq w, 0 \leq y \leq c\}, \tag{1.1}$$
$$Q(A, w, c) = \{y : y A \leq w, 0 \leq y \leq c\}. \tag{1.2}$$

* The work of this author was supported in part by N.S.F. Grant GP-32316X and by O.N.R. Grant N00014-67A-0077-002F.
** The work of this author was supported in part by the U.S. Army under contract #DAHCO4-C-0023.

Let $1 = (1, \ldots, 1)$ denote the vector all of whose components are 1. The two questions are:

> If $P(A, w, c)$ is not empty, is the minimum value of $1 \cdot y$, taken over all $y \in P(A, w, c)$, achieved at an integral vector y? (1.3)

> Is the maximum value of $1 \cdot y$, taken over all $y \in Q$ (A, w, c) achieved at an integral vector y? (1.4)

Berge defines a $(0,1)$-matrix A to be balanced if A contains no square submatrix of odd order whose row and column sums are all two. He shows that the answer to (1.3) is affirmative for all $(0,1)$-vectors w and c if and only if A is balanced. He shows that the answer to (1.4) is affirmative for all w whose components are 1 or ∞ and for all $(0,1)$-vectors c if and only if A is balanced. Finally, he remarks that for all c whose components are 0 or ∞ and all w whose components are nonnegative integers, the Lovász-Fulkerson perfect graph theorem [4, 6, 7] implies that the answer to (1.3) is affirmative if and only if A is balanced.

In this paper we prove that if A is balanced, then the answers to (1.3) and (1.4) are affirmative for all nonnegative integral w and c. We do not use the perfect graph theorem as a lemma, nor the results of Berge in [2] or in earlier work on balanced matrices [1].

The above results and those of Berge are used to relate the theory of balanced matrices to those of blocking pairs of matrices and anti-blocking pairs of matrices [3, 4, 5]. We summarize below some pertinent aspects of these two geometric duality theories.

We first discuss briefly the blocking theory. Let A be a nonnegative m by n matrix, and consider the convex polyhedron

$$\{x : A x \geq 1, x \geq 0\}. \tag{1.5}$$

A row vector a^i of matrix A is inessential (does not represent a facet of (1.5)) if and only if a^i is greater than or equal to a convex combination of other rows of A. The (nonnegative) matrix A is proper if none of its rows is inessential. Let A be proper with rows $a^1, \ldots, a^m$. Let B be the r by n matrix having rows $b^1, \ldots, b^r$, where $b^1, \ldots, b^r$ are the extreme points of (1.5). Then B is proper and the extreme points of the polyhedron

$$\{x : B x \geq 1, x \geq 0\} \tag{1.6}$$

are $a^1, \ldots, a^m$. The matrix B is called the blocking matrix of A and vice-versa. Together, A and B constitute a blocking pair of matrices, and the polyhedra (1.5) and (1.6) they generate are called a blocking pair of polyhedra. (Thus for any blocking pair of polyhedra, the non-trivial facets of one and the extreme points of the other are represented by exactly the same vectors; trivial facets are those corresponding to the nonnegativity constraints.)

Let A be a nonnegative m by n matrix and consider the packing program

$$
\begin{aligned}
\text{maximize} \quad & 1 \cdot y, \\
\text{subject to} \quad & y\,A \leq w, \qquad y \geq 0,
\end{aligned}
\tag{1.7}
$$

where w is nonnegative. Let B be an r by n nonnegative matrix having rows $b^1, \ldots, b^r$. The max–min equality is said to hold for the ordered pair A, B if, for every n-vector $w \geq 0$, the packing program (1.7) has a solution vector y such that

$$
1 \cdot y = \min_{1 \leq j \leq r} b^j \cdot w.
\tag{1.8}
$$

One theorem about blocking pairs asserts that the max–min equality holds for the ordered pair of proper matrices A, B if and only if A and B are a blocking pair. Hence if the max–min equality holds for A, B, it also holds for B, A. (Note that the addition of inessential rows to either A or B does not affect the max–min equality.)

Now let A be a proper (0,1)-matrix, with blocking matrix B. The strong max–min equality is said to hold for A, B if, for any nonnegative integral vector w, the packing program (1.7) has an integral solution vector y, which of course satisfies (1.8). A necessary, but not sufficient, condition for the strong max–min equality to hold for A, B is that each row of B be a (0,1)-vector. To say that an m by n (0,1)-matrix A is proper is simply to say that A is the incidence matrix of m pairwise non-comparable subsets of an n-set, i.e., A is the incidence matrix of a clutter. If the strong max–min equality holds for A and its blocking matrix B, then B is the incidence matrix of the blocking clutter, i.e., B has as its rows all (0,1)-vectors that make inner product at least 1 with all rows of A, and that are minimal with respect to this property. If A and B are a blocking pair of (0,1)-matrices, the strong max–min equality may hold for A, B, but need not hold for B, A. This is in decided contrast with the similar situation for anti-blocking pairs of matrices, which we next briefly discuss.

Let A be an m by n nonnegative matrix with rows $a^1, \ldots, a^m$, having no zero columns, and consider the convex polyhedron

$$\{x : A x \leq 1, x \geq 0\}. \tag{1.9}$$

(While a row vector a^i of A is inessential in (1.9) if and only if a^i is less than or equal to a convex combination of other rows of A, we shall not limit A to "proper" matrices in this discussion, as we did for blocking pairs, because there will not be a one–one correspondence between non-trivial facets of one member of a pair of anti-blocking polyhedra and the extreme points of the other.) Let D be the r by n matrix having rows $d^1, \ldots, d^r$, where $d^1, \ldots, d^r$ are the extreme points of (1.9). Then D is nonnegative, has no zero columns, and the extreme points of

$$\{x : D x \leq 1, x \geq 0\} \tag{1.10}$$

are $a^1, \ldots, a^m$ and all projections of $a^1, \ldots, a^m$. D is called an anti-blocking matrix of A, and vice-versa. Together, A and D constitute an anti-blocking pair of matrices, and the polyhedra (1.9) and (1.10) are an anti-blocking pair of polyhedra.

Now consider the covering program

$$\begin{aligned} \text{minimize} \quad & 1 \cdot y, \\ \text{subject to} \quad & y A \geq w, \quad y \geq 0, \end{aligned} \tag{1.11}$$

where w is nonnegative. Let D be an r by n nonnegative matrix having no zero columns with rows $d^1, \ldots, d^r$. The min–max equality is said to hold for the ordered pair A, D if, for every n-vector $w \geq 0$, the covering program (1.11) has a solution vector y satisfying

$$1 \cdot y = \max_{1 \leq j \leq r} d^j \cdot w. \tag{1.12}$$

Then the min–max equality holds for A, D if and only if A and D are an anti-blocking pair. Hence, if the min–max equality holds for A, D, it also holds for D, A.

Now let A be a (0,1)-matrix, with anti-blocking matrix D. The strong min–max equality is said to hold for A, D if, for every nonnegative integral vector w, the covering program (1.11) has an integral solution vector y; y of course satisfies (1.12). A necessary and sufficient condition for the strong min–max equality to hold for A, D is that all the essential rows of D be (0,1)-vectors. Hence, if the strong min–max equality holds for A, D, it also holds in the reverse direction D, A (where we may limit D to its essential rows). In this case, it can be shown that the essential (maximal) rows of A are the incidence vectors of the cliques of a graph G on n vertices, and the essential rows of D are the incidence vectors of the anti-cliques (maximal independent sets of vertices) of G. Graph G is thus pluperfect, or equivalently, perfect. The fact that the strong min–max equality for A, D implies the strong min–max equality for D, A is the essential content of the perfect graph theorem.

We shall show in Section 5 that the results described above and those of Berge imply: (a) If A is balanced and B is the blocking matrix of A, then the strong max–min equality holds for both A, B and B, A, and (b) if A is balanced and if D is an anti-blocking matrix of A, then the strong min–max equality holds for A, D (and hence for D, A).

2. Vertices of some polyhedra

We first state the lemmas of this section, and then give their proofs.

Lemma 2.1. *If A is balanced, and if $\{x : A x = 1, x \geq 0\}$ is not empty, then every vertex of this polyhedron has all coordinates 0 or 1.*

Lemma 2.2. *If A is balanced, and if $\{x : A x \geq 1, x \geq 0\}$ is not empty, then every vertex of this polyhedron has all coordinates 0 or 1.*

Lemma 2.3. *If A is balanced, and if $\{x, z : A x - z = 1, x \geq 0, z \geq 0\}$ is not empty, then every vertex of this polyhedron has all coordinates 0 or 1.*

Lemma 2.4. *If A is balanced, then every vertex of $\{x, z : A x - z \leq 1, x \geq 0, z \geq 0\}$ is integral. Hence if A is balanced, every vertex of $\{x : A x \leq 1, x \geq 0\}$ has coordinates 0 or 1.*

Note that Lemma 2.1 is a special case of Lemma 2.3, but it is convenient to separate the proofs.

Proof of Lemma 2.1. If A is balanced, then every submatrix of A is balanced. We shall prove Lemma 2.1 by induction on the number of rows of A. It is clearly equivalent to prove that if $x > 0$ satisfies $A x = 1$, then there exists a set of non-overlapping columns $a_{j_1}, \ldots, a_{j_k}$ of A (i.e., $a_{j_r} \cdot a_{j_s} = 0$ for $r \neq s$) whose sum is the vector 1. For any set S of non-overlapping columns, define $C(S)$, the "cover of S", to be the number of i such that $\sum_{j \in S} a_{ij} = 1$. Let S^* be a set of non-overlapping columns such that $C(S^*) \geq C(S)$ for any set S of non-overlapping columns. If $C(S^*) = m =$ number of rows of A, we are done, so assume $C(S^*) = k < m$, and, say, $\sum_{j \in S_*} a_{ij} = 1$ for $i = 1, \ldots, k$. Let $\bar{A}$ be the submatrix of A formed by rows $1, \ldots, k$. We have $\bar{A} x = 1$, $x > 0$, so, by the induction hypothesis, any column of $\bar{A}$ is contained in a set T of non-overlapping columns of $\bar{A}$ such that $C(T) = k$. In particular, let j^* be a column index such that $a_{ij}^* = 1$ for some $i \in \{k + 1, \ldots, m\}$, and let the aforementioned T contain j^*. Now some column indices in T (possibly none) may coincide with some column indices in S^*. Let $V = T \setminus S^*$, $U = S^* \setminus T$, both non-empty. Define a graph $G(\bar{A})$ whose points are the indices in $V \cup U$, with j and ℓ adjacent if and only if $\bar{a}_j \cdot \bar{a}_\ell > 0$. Clearly, $G(A)$ is bipartite with parts U and V. Let W be the vertices of the connected component $W(\bar{A})$ of $G(\bar{A})$ containing j^* (W may be $V \cup U$). It follows that

$$\sum_{j \in U \cap W} a_{ij} = \sum_{j \in V \cap W} a_{ij} = 0 \text{ or } 1 \quad \text{for } i = 1, \ldots, k. \tag{2.1}$$

Suppose that, for each $i = k + 1, \ldots, m$,

$$\sum_{j \in V \cap W} a_{ij} \leq 1. \tag{2.2}$$

Since $j^* \in W$, it follows from (2.1) and (2.2) that the columns of A with indices in

$$(S^* \setminus (U \cap W)) \cup (V \cap W)$$

are a non-overlapping set of columns with cover $\geq k + 1$, contradicting the definition of S^*. Hence (2.2) is untenable. Now consider the graph $W(A)$ with point set W, where j and ℓ are adjacent if and only if $a_j \cdot a_\ell > 0$. Recall that $W(\bar{A})$ is connected and bipartite. The graphs $W(\bar{A})$ and $W(A)$ have the same point set W, but $W(A)$ has more edges. In particular, there exists at least one pair of points in $W \cap V$ which are adjacent in $W(A)$. Let j and ℓ be points in $W \cap V$ such that the shortest path P in $W(\bar{A})$ joining j and ℓ contains no points j' and ℓ' in $W \cap V$ adjacent in $W(A)$ other than j and ℓ. Clearly such a path exists and is of even length. Let this path be

$$j = j_1, i_1, j_2, i_2, \ldots, j_p, i_p, j_{p+1} = \ell,$$

where the first, third, fifth, ... indices are in V, the second, fourth, ... indices are in U. Let $r^* \in \{k + 1, \ldots, m\}$ satisfy $a_{r^* j_1} = a_{r^* j_{p+1}} = 1$ and choose $r_1, \ldots, r_p, s_1, \ldots, s_p$ such that

$$a_{r_t j_t} = a_{r_t i_t} = 1, \qquad t = 1, \ldots, p,$$
$$a_{s_t i_t} = a_{s_t j_{t+1}} = 1, \qquad t = 1, \ldots, p.$$

That such indices exist follows from the construction of the path P. It is now clear that the submatrix of A formed by the columns $i_1, \ldots,$ $i_p, j_1, \ldots, j_{p+1}^{\cdot}$ and rows $r^*, r_1, \ldots, r_p, s_1, \ldots, s_p$ violates the hypothesis that A is balanced. Thus $C(S^*) = m$, proving Lemma 2.1.

Proof of Lemma 2.2. If x is a vertex of $\{x : A x \geq 1, x \geq 0\}$, it is a vertex of the polyhedron obtained by deleting the inequalities of $A x \geq 1$ that are strict. By Lemma 2.1, every vertex of this polyhedron has all coordinates 0 or 1.

Proof of Lemma 2.3. If (x, z) is a vertex of $\{x, z : A x - z = 1, x \geq 0,$ $z \geq 0\}$, then x is a vertex of $\{x : A x \geq 1, x \geq 0\}$. Lemma 2.3 thus follows from Lemma 2.2.

Proof of Lemma 2.4. If (x, z) is a vertex of $\{x, z : A x - z \leq 1, x \geq 0,$ $z \geq 0\}$, it is a vertex of the polyhedron obtained by deleting the inequalities of $A x - z \leq 1$ that are strict. Thus Lemma 2.4 follows from Lemma 2.3.

3. Solution of Problem (1.3)

We first prove a lemma.

Lemma 3.1. *Let A be a $(0,1)$-matrix satisfying the condition: For all nonnegative integral vectors w and c such that $P(A, w, c)$ is not empty, the minimum value of $1 \cdot y, y \in P(A, w, c)$ is an integer. Then for all nonnegative integral vectors w and c such that $P(A, w, c)$ is not empty, there exists an integral vector y that minimizes $1 \cdot y$ over $y \in P(A, w, c)$.*

Proof. The lemma is true if $1 \cdot c = 0$, and so we argue by induction on $1 \cdot c$.

Assume $y = (y_1, y_2, \ldots, y_m)$ is a solution to the linear program

$$
\begin{aligned}
\text{minimize} \quad & 1 \cdot y, \\
\text{subject to} \quad & y \in P(A, w, c),
\end{aligned}
\tag{3.1}
$$

with at least one component not integral, say $y_1 = r + \theta$, where $r \geq 0$ is an integer and $0 < \theta < 1$. Let $1 \cdot y = k$, where k is an integer. For any number z, define $z^+ = \max(0, z)$, and for any vector $z = (z_1, z_2, \ldots)$, define $z^+ = (z_1^+, z_2^+, \ldots)$. Let $\alpha = (r, y_2, \ldots, y_m)$, and note that $0 \leq \alpha \leq \tilde{c} = (c_1 - 1, c_2, \ldots, c_m)$. Let a^1 be the first row of A. Since $\alpha A \geq w - a^1$ and $\alpha A \geq 0$, we have $\alpha A \geq (w - a^1)^+$. Thus $\alpha \in P(A, (w - a^1)^+, \tilde{c})$, and $1 \cdot \alpha = k - \theta < k$. Now $1 \cdot \tilde{c} < 1 \cdot c$. Hence, by the induction assumption there exists an integral vector $\beta = (\beta_1, \ldots, \beta_m)$ such that $\beta A \geq (w - a^1)^+ \geq w - a^1, 0 \leq \beta \leq \tilde{c}$, and $1 \cdot \beta = \ell \leq k - \theta < k$, where ℓ is an integer. Therefore, the integral vector $\bar{\beta} = (\beta_1 + 1, \beta_2, \ldots, \beta_m) \in P(A, w, c), 1 \cdot \bar{\beta} = \ell + 1 \leq k$. But no solution to (3.1) can have value less than k, and hence $1 \cdot \bar{\beta} = k$. Thus $\bar{\beta}$ is an integral vector solving (3.1).

Theorem 3.2.[1] *Let A be balanced, and let w and c be nonnegative integral vectors such that $P(A, w, c)$ is not empty. Then the linear program (3.1) has an integral solution.*

Proof. Since $P(A, w, c)$ is not empty and bounded, (3.1) has a solution. Hence, by the duality theorem of linear programming, the dual program

$$
\begin{aligned}
\text{maximize} \quad & w \cdot x - c \cdot z, \\
\text{subject to} \quad & A x - z \leq 1, \qquad x \geq 0, \qquad z \geq 0,
\end{aligned}
\tag{3.2}
$$

[1] Added in proof: A different (and earlier) demonstration of Theorem 3.2 was given by L. Lovász.

has a solution. One such must occur at a vector with integral coordinates, by Lemma 2.4, so the common value of (3.2) and of (3.1) is an integer. But this means that the hypothesis of Lemma 3.1 holds. Hence, the conclusion of Lemma 3.1 holds, proving the theorem.

Note that the theorem holds if all coordinates of the vector c are ∞, an observation we will need below.

4. Solution of Problem (1.4)

We devote this section to the proof of Theorem 4.1 below.

Theorem 4.1. *Let A be a balanced matrix, and let w and c be nonnegative integral vectors. Then the linear program*

$$\begin{aligned} \text{maximize} \quad & 1 \cdot y, \\ \text{subject to} \quad & y \in Q(A, w, c) \end{aligned} \tag{4.1}$$

has an integral solution vector y.

Proof. We first remark that if A is balanced, the matrix (A, I) is balanced. Thus it suffices to prove that if A is balanced and $w \geq 0$ is integral, then the linear program

$$\begin{aligned} \text{maximize} \quad & 1 \cdot y, \\ \text{subject to} \quad & y A \leq w, \qquad y \geq 0, \end{aligned} \tag{4.2}$$

has an integral solution vector y. We shall prove this by a double induction on the pair of integers $(1 \cdot w, m)$, where A has m rows. Note that the theorem clearly is valid for any $m \geq 1$ if $1 \cdot w = 0$; it is also valid for any nonnegative integer value of $1 \cdot w$, if $m = 1$ (i.e., if (4.2) is a problem in one variable.)

Let $y = (y_1, y_2, \ldots, y_m)$ be a fractional solution of (4.2). If at least one y_i is zero, we are in the situation described by the pair of integers $(1 \cdot w, m - 1)$, since any submatrix of A is balanced, and the induction hypothesis applies. Thus we suppose all $y_i > 0$. By Lemma 2.2 and the duality theorem of linear programming, we know that $1 \cdot y = k$, where k is an integer. Now suppose there is at least one j such that $y \cdot a_j < w_j$, where a_j is the jth column of A. Thus $w_j > 0$. If $y \cdot a_j \leq w_j - 1$, we consider the pair of integers $(1 \cdot w - 1, m)$. By the inductive hypothesis, there is an integral vector z such that $z A \leq 0$, $z \geq 0$, $1 \cdot z = 1 \cdot y = k$, and we are done. Thus we may assume that $y \cdot a_j = w_j - 1 + \theta$, where $0 < \theta < 1$. Hence $a_j \neq 0$. Then clearly we can find a vector z such that $z \geq 0$, $z A \leq (w_1, w_2, \ldots, w_j - 1, \ldots, w_n)$, $z \leq y$, and $1 \cdot z = k - \theta$. By the inductive hypothesis for the pair of integers $(1 \cdot w - 1, m)$, there is an integral vector α satisfying $\alpha \geq 0$, $\alpha A \leq (w_1, \ldots, w_j - 1, \ldots, w_n) \leq w$, $1 \cdot \alpha \geq k - \theta$, hence $1 \cdot \alpha = k$, and we are done.

Thus $y \cdot a_j = w_j$ for all j and $y_i > 0$ for all i. By the principle of complementary slackness, every optimal solution of the dual problem

$$
\begin{aligned}
&\text{minimize} &&w \cdot x, \\
&\text{subject to} &&A x \geq 1, &&x \geq 0
\end{aligned}
\tag{4.3}
$$

satisfies $A x = 1$, $x \geq 0$, $w \cdot x = k$. Select one such x. Then y and x are optimal solutions, respectively, of the dual programs

$$
\begin{aligned}
&\text{minimize} &&1 \cdot y, \\
&\text{subject to} &&y A \geq w, &&y \geq 0,
\end{aligned}
\tag{4.4}
$$

$$
\begin{aligned}
&\text{maximize} &&w \cdot x, \\
&\text{subject to} &&A x \leq 1, &&x \geq 0,
\end{aligned}
\tag{4.5}
$$

with common value $1 \cdot y = w \cdot x = k$. By the remark at the end of the last section, there exists an integral vector α such that $\alpha \geq 0$, $\alpha A \geq w$, $1 \cdot \alpha = k$. If $\alpha A = w$, we are done. So assume $\alpha \cdot a_j > w_j$ for at least one j. Since $y_i > 0$ for all i, there is a number t, $0 < t < 1$, such that $y_i > (1 - t)\alpha_i$ for all i. Let vector z solve $y = (1 - t)\alpha + t z$, i.e., $z = (1/t) [y - (1 - t)\alpha]$. Thus $z \geq 0$ and $1 \cdot z = k$. Now, since $y A = w$ and $\alpha A \geq w$, it follows that $z A \leq w$. Moreover, since there is a j such that $\alpha \cdot a_j > w_j$, we have $z \cdot a_j < w_j$. Thus z is a solution to (4.1) with $z \cdot a_j < w_j$ for some j. However, as we have already seen, in this case the theorem is true by induction, and this completes the proof of Theorem 4.1.

5. Blocking pairs and anti-blocking pairs

Our purpose in this section is to prove the following theorems, which were mentioned in Section 1.

Theorem 5.1. *Let A be balanced and let B be the blocking matrix of A. Then the strong max–min equality holds for both A, B and B, A.*

Theorem 5.2. *Let A be balanced with no zero columns and let D be an anti-blocking matrix of A. Then the strong min–max equality holds for both A, D and D, A.*

Note that we have not assumed the $(0,1)$-matrix A in the statement of Theorem 5.1 to be proper; it would be no restriction to do so, however; we could just consider the minimal (essential, in the blocking sense) rows of A.

Proof of Theorem 5.1. That the strong max–min equality holds for the ordered pair A, B follows from Theorem 4.1 by taking the components of the vector c in Theorem 4.1 all equal to ∞.

To show that the strong max–min equality holds in the reverse direction B, A, we first note that [2, Theorem 2] can be rephrased in blocking terminology as follows: Let A be balanced and let B have as its rows all $(0,1)$-vectors that make inner product at least 1 with every row of A and that are minimal with respect to this property (i.e., B is the incidence matrix of the blocking clutter of the clutter of minimal rows of A); then the linear program

$$\begin{aligned} \text{maximize} \quad & 1 \cdot y, \\ \text{subject to} \quad & y\,B \leq 1, \qquad y \geq 0, \end{aligned} \tag{5.1}$$

has a $(0,1)$ solution vector y satisfying $1 \cdot y = \min 1 \cdot a^i$, taken over all rows a^i of A. To get the strong max–min equality for B, A from this, we need to pass from the vector 1 on the right-hand side of $y\,B \leq 1$ to a general nonnegative integral vector w. This transformation can be effected inductively by first observing that if A is balanced, and if we duplicate a column of A, the resulting matrix A' is balanced [2, Prop. 5]. Pictorially:

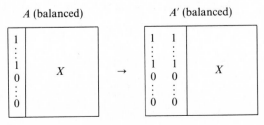

B (blocker of A)   B' (blocker of A')

Thus, if the first component of w is 2 instead of 1, we can consider the linear program

$$\text{maximize} \quad 1 \cdot y,$$
$$\text{subject to} \quad y\,B' \le 1, \qquad y \ge 0 \tag{5.2}$$

instead of

$$\text{maximize} \quad 1 \cdot y,$$
$$\text{subject to} \quad y\,B \le (2, 1, \ldots, 1), \qquad y \ge 0. \tag{5.3}$$

It follows that a general nonnegative integral vector w can be dealt with by deleting certain columns of A (those corresponding to zero components of w), replicating others, yielding a new balanced matrix, and making the appropriate transformations on the blocker B of A (a zero component of w means that we delete the corresponding column of B and also delete all rows of B that had a 1 in that column). In this way, one can deduce from [2, Theorem 2] that if A is balanced, the strong max–min equality holds for B, A.

In connection with Theorem 5.1 and its proof, we point out that the blocking matrix B of a balanced matrix A may not be balanced. For example, let

$$A = \begin{bmatrix} 1 & 1 & 0 & 0 & 0 & 0 & 0 \\ 0 & 0 & 1 & 1 & 0 & 0 & 0 \\ 0 & 0 & 1 & 0 & 0 & 0 & 1 \\ 0 & 0 & 0 & 1 & 0 & 1 & 0 \\ 0 & 0 & 0 & 0 & 1 & 1 & 1 \end{bmatrix}$$

Matrix A is balanced, with blocking matrix

$$B = \begin{bmatrix} 0 & 1 & 1 & 1 & 1 & 0 & 0 \\ 0 & 1 & 1 & 0 & 0 & 1 & 0 \\ 0 & 1 & 0 & 1 & 0 & 0 & 1 \\ 1 & 0 & 1 & 1 & 1 & 0 & 0 \\ 1 & 0 & 1 & 0 & 0 & 1 & 0 \\ 1 & 0 & 0 & 1 & 0 & 0 & 1 \end{bmatrix}$$

Proof of Theorem 5.2. If the (0,1)-matrix A has no zero columns, then $P(A, w, c)$ is not empty, where c is the vector all of whose components are ∞. The strong min–max equality for A, D, where D is an anti-blocking matrix of A, now follows from Theorem 3.2 and the discussion in Section 1 concerning anti-blocking pairs. Moreover, as noted in Section 1, the strong min–max equality for A, D implies the strong min–max equality for D, A.

Theorem 5.2 can be paraphrased as follows. The maximal (essential, in the anti-blocking sense) rows of a balanced matrix A are the incidence vectors of the cliques of a perfect graph G. Consequently, the essential rows of D are the incidence vectors of the anti-cliques of G.

References

[1] C. Berge, *Graphes et hypergraphes* (Dunod, Paris, 1970) ch. 20.
[2] C. Berge, "Balanced matrices", *Mathematical Programming* 2 (1972) 19–31.
[3] D.R. Fulkerson, "Blocking polyhedra", in: *Graph theory and its applications*, Ed. B. Harris (Academic Press, New York, 1970) pp. 93-112.
[4] D.R. Fulkerson, "Anti-blocking polyhedra", *Journal of Combinatorial Theory* 12 (1) (1972) 50–71.
[5] D.R. Fulkerson, "Blocking and anti-blocking pairs of polyhedra", *Mathematical Programming* 1 (1971) 168–194.
[6] D.R. Fulkerson, "On the perfect graph theorem", in: *Mathematical programming*, Eds. T.C. Hu and S.M. Robinson (Academic Press, New York, 1973) pp. 69–76.
[7] L. Lovász, "Normal hypergraphs and the perfect graph conjecture", *Discrete Mathematics* 2 (1972) 253–267.

Mathematical Programming Study 1 (1974) 133–147. North-Holland Publishing Company

DERIVATION OF A BOUND FOR ERROR-CORRECTING CODES USING PIVOTING TECHNIQUES

James H. GRIESMER

IBM Thomas J. Watson Research Center, Yorktown Heights, New York, U.S.A.

Received 31 January 1974

Dedicated to Professor A.W. Tucker in gratitude for his guidance

Using an integer programming formulation of the problem of finding a binary linear error-correcting code of minimum length n with 2^k code words and minimum distance d, a lower bound on n is constructed. The bound is shown to be equivalent to one previously obtained. Possible extensions using this approach are indicated.

1. Introduction

We briefly sketch the notions from coding theory which are needed to provide a background for the results contained in this paper. A more complete discussion of the field may be obtained from such references as the book by Peterson and Weldon [11], or the book chapter by Solomon [13].

We are concerned with the problem of transmitting messages over a noisy communications channel. We shall assume messages consist of words, or sequences, all of the same length, n, having components which are either 0 or 1. The channel we are considering is known as the (memoryless) *binary symmetric channel*, where the probability of a 0 becoming a 1 is equal to that of a 1 becoming a 0.

A code will be taken to be a subset of the 2^n possible messages or code words of length n, which are distinguishable from each other under reasonable noise conditions. The concept of *Hamming distance* is used as the distinguishability criterion. Two code words are said to have Hamming distance d if they differ in precisely d out of n positions. When a message is received, we shall assume that the fewest errors possible occurred in transmission, and search for the closest code word under the Hamming distance criterion. If the code is chosen so that the distance between any two code words is at least $d = 2e + 1$, then the code is e-error-correcting.

We add the requirement that such fixed-length codes be *group* or *linear codes*; i.e., subspaces of the vector space of n-tuples with components from GF(2). Such a linear code will have dimension k or, equivalently, 2^k code words, for some $k \leq n$, and will be referred to as an (n, k) *linear code*.

The *weight* of a code word is defined to be the number of its components which are equal to 1. Clearly the distance between two code words is equal to the weight of that code word which is their sum. Hence, for an (n, k) linear code, we can ensure that the code is e-error-correcting, by requiring that the weight of each of the $2^k - 1$ non-zero code words is at least $d = 2e + 1$.

A compact representation of an (n, k) linear code is obtained through the use of a set·of k basis vectors for the k-dimensional subspace. These form a $k \times n$ *generator matrix*. All code words are then linear combinations of the rows of the generator matrix. A particular generator matrix will consist of a set of columns taken from the collection of all $2^k - 1$ non-zero k-tuples of binary digits. Within a permutation of the columns, a $k \times n$ generator matrix, and hence, an (n, k) linear code, can be described by a list of the number of columns of each type. Such a description is called the *modular representation* [12] of the code: a vector of $2^k - 1$ non-negative integers,

$$N = (n_1, n_2, \ldots, n_{2^k-1}),$$

where n_i is the number of columns of type i (a column is of type i if it is the binary representation of the integer i), and

$$\sum_{i=1}^{2^k-1} n_i = n.$$

The weights of non-zero-code words in the corresponding (n, k) linear code may be obtained by multiplying the vector N by a $(2^k - 1) \times (2^k - 1)$ matrix C_k:

$$W = C_k N^\mathrm{T}.$$

Here, W^T is a $(2^k - 1)$-component vector:

$$W^T = (w_1, w_2, \ldots, w_{2k-1})$$

with w_i indicating the weight of that code word which is formed by combining the rows of the generator matrix which correspond to the occurrence of 1's in the binary representation of i. The entries of C_k are given by:

$$c_{i,j} = \begin{cases} 1 & \text{if } (i)_2 \text{ and } (j)_2 \text{ have an odd number of ones in common,} \\ 0 & \text{otherwise,} \end{cases}$$

where $(i)_2$ and $(j)_2$ denote the binary representation of i and j, respectively.

As an example, we use the following 3×5 matrix as the generator matrix of a $(5, 3)$ linear code:

$$\begin{bmatrix} 1 & 1 & 1 & 0 & 1 \\ 0 & 1 & 0 & 1 & 1 \\ 0 & 0 & 1 & 1 & 1 \end{bmatrix}.$$

The modular representation for this code is the vector $N = (1, 0, 1, 0, 1, 1, 1)$. The weights of the 7 non-zero code words are given by

$$W = C_3 N^T = \begin{bmatrix} 1 & 0 & 1 & 0 & 1 & 0 & 1 \\ 0 & 1 & 1 & 0 & 0 & 1 & 1 \\ 1 & 1 & 0 & 0 & 1 & 1 & 0 \\ 0 & 0 & 0 & 1 & 1 & 1 & 1 \\ 1 & 0 & 1 & 1 & 0 & 1 & 0 \\ 0 & 1 & 1 & 1 & 1 & 0 & 0 \\ 1 & 1 & 0 & 1 & 0 & 0 & 1 \end{bmatrix} \begin{bmatrix} 1 \\ 0 \\ 1 \\ 0 \\ 1 \\ 1 \\ 1 \end{bmatrix} = \begin{bmatrix} 4 \\ 3 \\ 3 \\ 3 \\ 3 \\ 2 \\ 2 \end{bmatrix}$$

The first, second, and fourth components of the weight vector are the weights of the first, second, and third rows of the generator matrix, respectively. The third component of the weight vector is the weight of the non-zero code word formed by adding the first and second rows of the generator matrix, etc.

Our interest, of course, is in proceeding in the reverse direction. Given a (column) vector of weights W (in our case, all of its components equal to some desired minimum weight, d), we wish to find a vector N of non-negative integers, which specifies an (n, k) linear code whose non-zero code words have weights greater than or equal to d. That is, we wish to find a vector N with non-negative integer components which satisfies the inequality system:

$$C_k N^T \geqslant W.$$

Ideally, one wishes to achieve such a linear code with the property that n, the code word length, is minimized. We use $N(k, d)$ to denote the minimum value of n, for a given k and d.

Solution of the following integer program [9] will yield the value of $N(k, d)$, as well as the modular representation of an (n, k) linear code with $n = N(k, d)$:

$$\text{minimize} \quad n = \sum_{j=1}^{2^k-1} n_j,$$

$$\text{subject to} \quad \sum_{j=1}^{2^k-1} c_{ij} n_j \geqslant d, \qquad i = 1, \ldots, 2^k - 1,$$

$$n_j \geqslant 0, \text{integer}, \quad j = 1, \ldots, 2^k - 1.$$

(1)

In Sections 2 and 3, we apply the methods developed by Gomory in the integer programming algorithm described in [3], to generate additional inequalities which are satisfied by the integer solutions to (1). The additional inequalities are then combined to obtain a lower bound on $N(k, d)$ in Section 4. Section 5 contains a proof of the equivalence of this bound with one previously obtained [9]. A final section discusses related work and possible extensions.

2. Linear programming solution

We can solve (1) as a linear program, i.e., by requiring only that $n_j \geqslant 0$, not that n_j be an integer. The optimal solution is $n_j = d/2^{k-1}, j = 1, \ldots, 2^k - 1$, giving value $n = (2^k - 1) d/2^{k-1}$ to the objective function. That this is an optimal solution to (1), considered as a linear program, is immediate since the dual to (1)

$$\text{maximize} \quad u = d \cdot \sum_{i=1}^{2^k-1} u_i,$$

$$\text{subject to} \quad \sum_{i=1}^{2^k-1} u_i c_{ij} \geqslant 1, \qquad j = 1, \ldots, 2^k - 1,$$

$$u_i \geqslant 0, \qquad i = 1, \ldots, 2^k - 1 \tag{2}$$

has a feasible solution $u_i = 1/2^{k-1}$, $i = 1, \ldots, 2^k - 1$, giving value $u = (2^k - 1) \, d/2^{k-1}$ to its objective function.

To find the matrix or tableau which gives rise to the solution of the minimization problem, i.e., the matrix which is obtained from successive pivot steps of the simplex method, we consider the inequality system for (1) in matrix form:

$$C_k N^T \geqslant D^T, \tag{3}$$

where $N = (n_1, \ldots, n_{2^k-1})$ and $D = (d, \ldots, d)$. Introducing slack variables $T = (t_1, \ldots, t_{2^k-1})$, we obtain the equality system

$$C_k N^T - T^T = D^T. \tag{4}$$

Since C_k is non-singular (we compute its determinant in the next section), we obtain the equivalent system:

$$N^T = C_k^{-1} D^T + C_k^{-1} T^T. \tag{5}$$

$C_k^{-1} = [\hat{c}_{ij}]$ is given explicitly in [2]:

$$\hat{c}_{ij} = \begin{cases} 1/2^{k-1} & \text{if } (i)_2 \text{ and } (j)_2 \text{ have an odd number of ones} \\ & \text{in common,} \\ -1/2^{k-1} & \text{otherwise.} \end{cases}$$

When T has all its components set equal to 0, we have

$$N^T = C_k^{-1} D^T \tag{6}$$

which can be easily evaluated: $n_j = d/2^{k-1}, j = 1, \ldots, 2^k - 1$, the optimal solution to (1). Hence we can write out the final tableau using Tucker's condensed schemata [15, 16]:

	1	$-t_1$	$-t_2$	$\cdots$	$-t_{2^k-1}$
$-n$	$-(2^k-1)\,d/2^{k-1}$	$1/2^{k-1}$	$1/2^{k-1}$	$\cdots$	$1/2^{k-1}$
n_1	$d/2^{k-1}$				
n_2	$d/2^{k-1}$				
n_3	$d/2^{k-1}$		$-\hat{c}_{ij}$		
$\vdots$	$\vdots$				
n_{2^k-1}	$d/2^{k-1}$				

We remark that (7) represents the final tableau arising from the problem of *maximizing* $-n = \sum_{j=1}^{2^k-1}(-n_j)$, which, of course, is equivalent to the problem of *minimizing* $n = \sum_{j=1}^{2^k-1} n_j$.

As an example, we give the final tableau for the case $k = 3$:

	1	$-t_1$	$-t_2$	$-t_3$	$-t_4$	$-t_5$	$-t_6$	$-t_7$
$-n$	$-7d/4$	$1/4$	$1/4$	$1/4$	$1/4$	$1/4$	$1/4$	$1/4$
n_1	$d/4$	$-1/4$	$1/4$	$-1/4$	$1/4$	$-1/4$	$1/4$	$-1/4$
n_2	$d/4$	$1/4$	$-1/4$	$-1/4$	$1/4$	$1/4$	$-1/4$	$-1/4$
n_3	$d/4$	$-1/4$	$-1/4$	$1/4$	$1/4$	$-1/4$	$-1/4$	$1/4$
n_4	$d/4$	$1/4$	$1/4$	$1/4$	$-1/4$	$-1/4$	$-1/4$	$-1/4$
n_5	$d/4$	$-1/4$	$1/4$	$-1/4$	$-1/4$	$1/4$	$-1/4$	$1/4$
n_6	$d/4$	$1/4$	$-1/4$	$-1/4$	$-1/4$	$-1/4$	$1/4$	$1/4$
n_7	$d/4$	$-1/4$	$-1/4$	$1/4$	$-1/4$	$1/4$	$1/4$	$-1/4$

3. The group of inequalities

We follow [3] in using the tableau (7) to derive additional inequalities which are satisfied by the non-negative integer solutions to (1). In general, if

$$x_i = a_{i0} + \sum_{j=1}^{m} a_{ij}(-t_j) \tag{8}$$

is an equation corresponding to one row of the tableau, then the inequality

$$0 \leqslant -f_{i0} - \sum_{j=1}^{m} f_{ij}(-t_j) \tag{9}$$

or the equivalent equation

$$s = -f_{i0} - \sum_{j=1}^{m} f_{ij}(-t_j) \tag{10}$$

is satisfied by the non-negative integer solutions to the original problem, where $0 \leqslant f_{ij} = a_{ij} - [a_{ij}], j = 0, \ldots, m$. In [3], the inequality (9) is also represented by an $(m + 1)$-component "fractional part" row vector

$$(f_{i0}, f_{i1}, f_{i2}, \ldots, f_{im}). \tag{11}$$

Under the operation of adding integer combinations of these fractional part rows, and reducing components modulo 1, the resulting vectors form a finite additive group of inequalities which are satisfied by the non-negative integer solutions to the original problem. The order of the group is the product of the pivot elements which were used in reaching the final tableau. (See [3] for details.)

In our special case, we have two types of fractional part rows derivable from the tableau (7):

$$G_0 = (f(-(2^k - 1)\,d/2^{k-1}), 1/2^{k-1}, 1/2^{k-1}, \ldots, 1/2^{k-1}),$$
$$G_i = (f(d/2^{k-1}), f_{i1}, f_{i2}, \ldots, f_{i2^k-1}), \qquad i = 1, \ldots, 2^k - 1,$$

where $f(-(2^k - 1)\,d/2^{k-1})$ and $f(d/2^{k-1})$ denote the fractional parts of $-(2^k - 1)\,d/2^{k-1}$ and $d/2^{k-1}$, respectively, and $f_{ij} = f(-\hat{c}_{ij})$, i.e.,

$$f_{ij} = \begin{cases} (2^{k-1} - 1)/2^{k-1} & \text{if } (i)_2 \text{ and } (j)_2 \text{ have an odd number} \\ & \text{of ones in common,} \\ 1/2^{k-1} & \text{otherwise.} \end{cases}$$

We show that $f(-(2^k - 1)\,d/2^{k-1}) = f(d/2^{k-1})$. Let $d = h\,2^{k-1} + r$, so that $d \equiv r \bmod 2^{k-1}$. Hence, $-2^k d + d \equiv r \bmod 2^{k-1}$ or $-(2^k - 1)d \equiv r \bmod 2^{k-1}$. Thus, $f(-(2^k - 1)\,d/2^{k-1}) = f(d/2^{k-1}) = r/2^{k-1}$.

Since the steps of the simplex method transform the matrix C_k into the matrix C_k^{-1}, the product of the pivot elements, and hence the order of the group generated by the fractional part rows, is simply the determinant of C_k. To evaluate it we partition C_k into submatrices:

$$
C_k = \left[\begin{array}{c|c|c}
C_{k-1} & \begin{matrix} 0 \\ \vdots \\ 0 \end{matrix} & C_{k-1} \\
\hline
0\ldots0 & 1 & 1\ldots1 \\
\hline
C_{k-1} & \begin{matrix} 1 \\ \vdots \\ 1 \end{matrix} & \overline{C}_{k-1}
\end{array} \right]
$$

where $\overline{C}_{k-1}$ is the matrix obtained from C_{k-1} by changing each 1 entry to a 0 and each 0 entry to a 1. We can perform a set of elementary row operations to obtain the following matrix which is equivalent to C_k and which has the same determinant:

$$
\left[\begin{array}{c|c|c}
C_{k-1} & \begin{matrix} 0 \\ \vdots \\ 0 \end{matrix} & C_{k-1} \\
\hline
0\ldots0 & 1 & 1\ldots1 \\
\hline
0 & \begin{matrix} 0 \\ \vdots \\ 0 \end{matrix} & -2C_{k-1}
\end{array} \right]
$$

Hence, we have the relation:

$$
|\det(C_k)| = 2^{2^{k-1}-1}(\det(C_{k-1}))^2. \tag{12}
$$

The following table gives values of $|\det(C_k)|$, and, hence, the order of the inequality group, for small values of k:

k	$|\det(C_k)|$
2	2
3	2^5
4	2^{17}
5	2^{49}
6	2^{129}

As an illustration, we list the 32 members of the group for the case $k = 3$:

$$G_0 = 1/4(4f(d/4), 1, 1, 1, 1, 1, 1, 1), \quad 2G_0 + G_0 = 1/4(4f(3d/4), 3, 3, 3, 3, 3, 3, 3),$$
$$G_1 = 1/4(4f(d/4), 3, 1, 3, 1, 3, 1, 3), \quad 2G_0 + G_1 = 1/4(4f(3d/4), 1, 3, 1, 3, 1, 3, 1),$$
$$G_2 = 1/4(4f(d/4), 1, 3, 3, 1, 1, 3, 3), \quad 2G_0 + G_2 = 1/4(4f(3d/4), 3, 1, 1, 3, 3, 1, 1),$$
$$G_3 = 1/4(4f(d/4), 3, 3, 1, 1, 3, 3, 1), \quad 2G_0 + G_3 = 1/4(4f(3d/4), 1, 1, 3, 3, 1, 1, 3),$$
$$G_4 = 1/4(4f(d/4), 1, 1, 1, 3, 3, 3, 3), \quad 2G_0 + G_4 = 1/4(4f(3d/4), 3, 3, 3, 1, 1, 1, 1),$$
$$G_5 = 1/4(4f(d/4), 3, 1, 3, 3, 1, 3, 1), \quad 2G_0 + G_5 = 1/4(4f(3d/4), 1, 3, 1, 1, 3, 1, 3),$$
$$G_6 = 1/4(4f(d/4), 1, 3, 3, 3, 3, 1, 1), \quad 2G_0 + G_6 = 1/4(4f(3d/4), 3, 1, 1, 1, 1, 3, 3),$$
$$G_7 = 1/4(4f(d/4), 3, 3, 1, 3, 1, 1, 3), \quad 2G_0 + G_7 = 1/4(4f/3d/4), 1, 1, 3, 1, 3, 3, 1),$$
$$G_0 + G_0 = 1/2(2f(d/2), 1, 1, 1, 1, 1, 1, 1), \quad 3G_0 + G_0 = (0, 0, 0, 0, 0, 0, 0, 0),$$
$$G_0 + G_1 = 1/2(2f(d/2), 0, 1, 0, 1, 0, 1, 0), \quad 3G_0 + G_1 = 1/4(0, 2, 0, 2, 0, 2, 0, 2),$$
$$G_0 + G_2 = 1/2(2f(d/2), 1, 0, 0, 1, 1, 0, 0), \quad 3G_0 + G_2 = 1/4(0, 0, 2, 2, 0, 0, 2, 2),$$
$$G_0 + G_3 = 1/2(2f(d/2), 0, 0, 1, 1, 0, 0, 1), \quad 3G_0 + G_3 = 1/4(0, 2, 2, 0, 0, 2, 2, 0),$$
$$G_0 + G_4 = 1/2(2f(d/2), 1, 1, 1, 0, 0, 0, 0), \quad 3G_0 + G_4 = 1/4(0, 0, 0, 0, 2, 2, 2, 2),$$
$$G_0 + G_5 = 1/2(2f(d/2), 0, 1, 0, 0, 1, 0, 1), \quad 3G_0 + G_5 = 1/4(0, 2, 0, 2, 2, 0, 2, 0),$$
$$G_0 + G_6 = 1/2(2f(d/2), 1, 0, 0, 0, 0, 1, 1), \quad 3G_0 + G_6 = 1/4(0, 0, 2, 2, 2, 2, 0, 0),$$
$$G_0 + G_7 = 1/2(2f(d/2), 0, 0, 1, 0, 1, 1, 0), \quad 3G_0 + G_7 = 1/4(0, 2, 2, 0, 2, 0, 0, 2).$$

4. The lower bound

We now combine subsets of the fractional part rows 2^{k-2} at a time to form new inequalities which are satisfied by non-negative integer solutions to (1). We define an operation on indices i and i', $1 \leqslant i, i' \leqslant 2^k - 1$:

$$(i \oplus i')_2 = (i)_2 \oplus (i')_2,$$

i.e., we consider $(i)_2$ and $(i')_2$ as k-dimensional vectors with components from GF (2), and add them as vectors. The result will be a k-dimensional vector which will be the binary representation of $i \oplus i'$.

We form the inequality:

$$G_0 + G_{i_1} + G_{i_2} + G_{i_3 = i_1 \oplus i_2} + G_{i_4} + G_{i_5 = i_1 \oplus i_4} \\ + \ldots + G_{i_{2^{k-3}}} + \ldots + G_{i_{2^{k-2}-1}}, \tag{13}$$

where $1 \leqslant i_p \leqslant 2^k - 1$, and $i_p \neq i_{p'}$ for $p \neq p'$, and where the addition of these fractional part row generators is carried out modulo 1. We show the resulting inequality has the following form:

$$(f(2^{k-2}d/2^{k-1}), 0, \ldots, 0, 1/2, 0, \ldots, 0, 1/2, 0, \ldots, 0, 1/2, 0, \ldots, 0). \tag{14}$$

The 0th component is 1/2 or 0, according to whether d is odd or even, since we are adding 2^{k-2} repetitions of $f(d/2^{k-1})$. The three positions where 1/2 entries occur are given by those indices j_1, j_2, and j_3 satisfying for all p, $1 \leqslant p \leqslant 2^{k-2} - 1$,

$$(j_r)_2 \perp (i_p)_2, \qquad r = 1, 2, 3,$$

considering the binary representations as elements of V_k, the vector space of k-tuples with components from GF(2). This is easy to see since the 2^{k-2} indices $(i_p)_2$ form a subspace S of dimension $k - 2$, and, hence, its orthogonal complement $S^\perp$ has dimension $k - (k - 2) = 2$, i.e., has 3 non-zero vectors, which we have called $(j_1)_2$, $(j_2)_2$, and $(j_3)_2$. Thus we have $(j_3)_2 = (j_1)_2 \oplus (j_2)_2$. Furthermore, any vector $(i)_2$ not in $S^\perp$ can be shown to be orthogonal to exactly half the elements of S. Considering the vectors $(i_1)_2, (i_2)_2, (i_4)_2, \ldots, (i_{2k-3})_2$ as a basis for S, let T_i be that subset of the basis which $(i)_2$ is orthogonal to and U_i be that subset of the basis which it is not. By assumption, $U_i \neq \emptyset$. Then $(i)_2$ will be orthogonal to all vectors formed by combining an even number of elements $u \in U_i$, with any number of elements of T_i, and not orthogonal otherwise. If $|U_i| = h > 0$ and $|T_i| = k - h - 2$, then the number of vectors in S orthogonal to $(i)_2$ is

$$2^{k-h-2} \left[\binom{h}{0} + \binom{h}{2} + \ldots + \begin{cases} \binom{h}{h-1} & h \text{ odd,} \\ \binom{h}{h} & h \text{ even} \end{cases} \right] = 2^{k-h-2} \cdot 2^{k-1} = 2^{k-3}.$$

Thus, in those components of the fractional part rows being added which are indexed by i, $i \neq j_1, j_2, j_3$, we are adding equal numbers of entries $(2^{k-1} - 1)/2^{k-1}$ and $1/2^{k-1}$, and reducing modulo 1, i.e., for such i, the corresponding component in the vector (14) is 0.

We now write out the inequality corresponding to (14):

$$0 \leqslant -f(2^{k-2}d/2^{k-1}) + 1/2 \, t_{j_1} + 1/2 \, t_{j_2} + 1/2 \, t_{j_3} \tag{15}$$

Now, for $r = 1, 2, 3$:

$$t_{j_r} = \sum_{j=1}^{2^k-1} c_{j_r j} n_j - d. \tag{16}$$

Substituting (16) into (15):

$$1/2 \sum_{j=1}^{2^k-1} c_{j_1,j} n_j + 1/2 \sum_{j=1}^{2^k-1} c_{j_2,j} n_j + 1/2 \sum_{j=1}^{2^k-1} c_{j_3,j} n_j \geq$$

$$\geq 3d/2 + f(2^{k-2}d/2^{k-1}). \tag{17}$$

Now $c_{j,j}$ is 1 if and only if j_r and j have an odd of ones in common, i.e., $(j_r)_2$ and $(j)_2$ are not orthogonal. Hence the only n_j not appearing in the sum are precisely those for which $(j)_2$ is orthogonal to $(j_1)_2$, $(j_2)_2$, and $(j_3)_2$, viz., $(i_1)_2, (i_2)_2, (i_3)_2, \ldots, (i_{2^{k-2}-1})_2$. All other n_j will appear twice in the sum, for if $j \neq i_p$, $p = 1, \ldots, 2^{k-2} - 1$, then $(j)_2$ will not be orthogonal to exactly two out of three of $(j_1)_2, (j_2)_2$ and $(j_3)_2$, since $(j_1)_2 \oplus (j_2)_2 = (j_3)_2$. Hence we have deduced the inequality

$$\sum_{j \neq i_1, i_2, \ldots, i_{2^k-2}-1} n_j \geq 3d/2 + f(2^{k-2}d/2^{k-1}) = I(\tfrac{3}{2}d) \tag{18}$$

where $I(\tfrac{3}{2}d)$ denotes "the least integer greater than or equal to $3d/2$."

We now fix $i_1, i_2, i_3 = i_1 \oplus i_2, \ldots, i_{2^{k-3}-1}$, and select a new index $\hat{i}_{2^{k-3}} \neq i_p$, $2^{k-3} \leq p \leq 2^{k-2} - 1$, plus all the indices $i_{2^{k-3}} \oplus i_p$, $1 \leq p \leq 2^{k-3} - 1$. We can now form another inequality of the form (13), which after substitution, yields another inequality of the form (18), in which the indices $i_1, i_2, \ldots, i_{2^{k-3}-1}$ are the same. Since we have $2^{k-3} - 1$ indices fixed out of a total of $2^k - 1$, we have left a set of $7 \cdot 2^{k-3}$ indices, and each time we pick an $i_{2^{k-3}}$, we use 2^{k-3} elements of this set. Hence we can form a total of 7 inequalities of the form (18) in which $i_1, i_2, \ldots, i_{2^{k-3}-1}$ are the same. Each of the variables n_j, $j \neq i_1, i_2, \ldots, i_{2^{k-3}-1}$, will appear in precisely 6 of the 7 inequalities. By adding these inequalities, we obtain the new inequality

$$\sum_{j \neq i_1, i_2, \ldots, i_{2^k-3}-1} n_j \geq 7 \, I(\tfrac{3}{2}d) \tag{19}$$

or

$$\sum_{j \neq i_1, i_2, \ldots, i_{2^k-3}-1} n_j \geq \tfrac{7}{6} I(\tfrac{3}{2}d). \tag{20}$$

Since the left-hand side of (20) is an integer, we may strengthen the right-hand side:

$$\sum_{j \neq i_1, i_2, \ldots, i_{2^{k-3}-1}} n_j \geq I(\tfrac{7}{6}I(\tfrac{3}{2}d)). \tag{21}$$

By a reasoning similar to the manner in which we formed 7 distinct inequalities of the form (18) in which $i_1, i_2, \ldots, i_{2^{k-3}-1}$ were kept fixed, we can now form 15 distinct inequalities of the form (21) in which $i_1, i_2, \ldots, i_{2^{k-4}-1}$ are kept fixed. Adding them and taking the integer part of the right-hand side, we obtain the inequality

$$\sum_{j \neq i_1, i_2, \ldots, i_{2^{k-4}-1}} n_j \geq I(\tfrac{15}{14}I(\tfrac{7}{6}I(\tfrac{3}{2}d))). \tag{22}$$

This process leads to the following chain of inequalities:

$$\sum_{j \neq i_1, i_2, \ldots, i_{2^{k-5}-1}} n_j \geq I(\tfrac{31}{30}I(\tfrac{15}{14}I(\tfrac{7}{6}I(\tfrac{3}{2}d)))), \tag{23}$$

$$\vdots$$

$$\sum_{j \neq i_1} n_j \geq I\left(\frac{2^{k-1}-1}{2^{k-1}-2}I\left(\frac{2^{k-2}-1}{2^{k-2}-2}I(\ldots I(\tfrac{31}{30}I(\tfrac{15}{14}I(\tfrac{7}{6}I(\tfrac{3}{2}d)))))\ldots))\right), \tag{24}$$

$$\sum_{j=1}^{2^k-1} n_j \geq I\left(\frac{2^k-1}{2^k-2}I\left(\frac{2^{k-1}-1}{2^{k-1}-2}I(\ldots I(\tfrac{31}{30}I(\tfrac{15}{14}I(\tfrac{7}{6}I(\tfrac{3}{2}d)))))\ldots))\right). \tag{25}$$

Hence we have proved

Theorem 1.

$$N(k, d) \geq I\left(\frac{2^k-1}{2^k-2}I\left(\frac{2^{k-1}-1}{2^{k-1}-2}I(\ldots I(\tfrac{31}{30}I(\tfrac{15}{14}I(\tfrac{7}{6}I(\tfrac{3}{2}d)))))\ldots))\right). \tag{26}$$

5. Equivalence of bounds

In [9], the following lower bound on $N(k, d)$ was derived.

$$N(k, d) \geq \sum_{i=0}^{k-1} [(d + 2^i - 1)/2^i] = \sum_{i=0}^{k-1} I(d/2^i), \tag{27}$$

where $[(d + 2^i - 1)/2^i]$ denotes the "greatest integer less than or equal to $(d + 2^i - 1)/2^i$." We now show that (26) and (27) are equivalent by proving

Theorem 2.

$$\sum_{i=0}^{k-1} I(d/2^i) = I\left(\frac{2^k - 1}{2^k - 2} I\left(\frac{2^{k-1} - 1}{2^{k-1} - 2} I(\ldots I(\tfrac{7}{6} I(\tfrac{3}{2} d))\ldots)\right)\right).$$
(28)

Proof. We use induction on k. Clearly, for $k = 2$, both the left- and right-hand sides of (28) are equal to $d + I(d/2)$. Assuming the theorem true for $k - 1$, we rewrite the right-hand side of (28):

$$I\left(\frac{2^k - 1}{2^k - 2} I\left(\frac{2^{k-1} - 1}{2^{k-1} - 2} I(\ldots I(\tfrac{7}{6} I(\tfrac{3}{2} d))\ldots)\right)\right) =$$

$$= I\left(\frac{2^{k-1} - 1}{2^{k-1} - 2} I(\ldots I(\tfrac{7}{6} I(\tfrac{3}{2} d))\ldots)\right)$$

$$+ I\left(\frac{1}{2^k - 2} I \frac{2^{k-1} - 1}{2^{k-1} - 2} I(\ldots I(\tfrac{7}{6} I(\tfrac{3}{2} d))\ldots)\right)$$

$$= \sum_{i=0}^{k-2} I(d/2^i) + I\left(\frac{1}{2^k - 2} I\left(\frac{2^{k-1} - 1}{2^{k-1} - 2} I(\ldots I(\tfrac{7}{6} I(\tfrac{3}{2} d))\ldots)\right)\right).$$

Hence it is sufficient to prove

$$I(d/2^{k-1}) = I\left(\frac{1}{2^k - 2} I\left(\frac{2^{k-1} - 1}{2^{k-1} - 2} I(\ldots I(\tfrac{7}{6} I(\tfrac{3}{2} d))\ldots)\right)\right).$$
(29)

Case A. $d = h\,2^{k-1}$, h an integer ≥ 0. Clearly, both left- and right-hand sides of (29) are equal to h.

Case B. $h\,2^{k-1} < h\,2^{k-1} + 1 \leq d < (h + 1)\,2^{k-1}$, h an integer ≥ 0. The left-hand side of (29) is $(h + 1)$ for any d satisfying the inequality. We show that the right-hand side of (29) is $(h + 1)$ when $d = h\,2^{k-1} + 1$.

$$I\left(\frac{1}{2^k - 2} I\left(\frac{2^{k-1} - 1}{2^{k-1} - 2} I(\ldots I(\tfrac{7}{6} I(\tfrac{3}{2}(h\,2^{k-1} + 1)))\ldots)\right)\right) =$$

$$= I\left(\frac{1}{2^k - 2} I\left(\frac{2^{k-1} - 1}{2^{k-1} - 2} I(\ldots I(\tfrac{7}{6}(3h\,2^{k-2} + 2))\ldots)\right)\right)$$

$$= I(\frac{1}{2^k - 2} I(\frac{2^{k-1} - 1}{2^{k-1} - 2} I(\ldots 7h\, 2^{k-3} + 3 \ldots)))$$

$$\vdots$$

$$= I(\frac{1}{2^k - 2}((2^{k-1} - 1)\, h\, 2 + (k - 1)))$$

$$= h + 1.$$

Since

$$h\, 2^{k-1} + 1 \leqslant d < (h + 1)2^{k-1}, \tag{30}$$

and since the two extremes of (30) make the right-hand side of (29) equal to $(h + 1)$, we conclude the right-hand side of (29) is equal to $(h + 1)$ for any intermediate value of d. This is immediate since the sense of the inequalities in (30) is preserved if all three parts are multiplied by the same positive scalar, or are operated on by the function $I(\ldots)$. But these are precisely the operations used in evaluating the right-hand side of (29).

6. Additional remarks

A considerably simplified proof of the bound obtained in [9] and in the present paper was given by Solomon and Stiffler [14]. They also generalized the bound to the case of a linear code over an arbitrary finite field GF(q). A paper by Baumert and McEliece [1] showed that this bound on $N(k, d)$ was sharp for a fixed k, when d was sufficiently large.

In [10], Helgert and Stinaff studied the function $d_{max}(n, k)$, the maximum minimum distance that can be attained by an (n, k) binary linear code, and tabulated upper and lower bounds on this function for values of $n \leqslant 127$ and $k \leqslant n$. They used the bound given in [9] in several cases to obtain values for the upper bound on $d_{max}(n, k)$.

The derivation of the lower bound on $N(k, d)$ using the techniques of integer programming is presented in the belief that additional research using such methods may lead to improved bounds for various values of d. It may be possible for example, to utilize the work of Gomory [4, 5, 6], and Gomory and Johnson [7, 8], to generate stronger inequalities than were possible following the approach used in this paper.

Acknowledgment

The author wishes to express his indebtedness to Professor A.W. Tucker for his guidance during the author's graduate study at Princeton University. The present paper has its roots in work done by the author during the IBM Research Institute on Combinatorial Problems held during the summer of 1959 under the leadership of Professor Tucker. The author would also like to thank Ralph Gomory and John Selfridge for helpful discussions on topics contained in this paper.

References

[1] L.D. Baumert and R.J. McEliece, "A note on the Griesmer bound", *IEEE Transactions on Information Theory* 19 (1973) 134–135.

[2] A.B. Fontaine and W.W. Peterson, "Group code equivalence and optimum codes", *IRE Transactions on Information Theory* 5 (1959) Special Supplement, pp. 60–70.

[3] R.E. Gomory, "An algorithm for integer solutions to linear programs", in: *Recent advances in mathematical programming*, Eds. R.L. Graves and P. Wolfe (McGraw-Hill, New York, 1963) pp. 269–302.

[4] R.E. Gomory, "Faces of an integer polyhedron", *Proceedings of the National Academy of Sciences* 57 (1967) 16–18.

[5] R.E. Gomory, "Some polyhedra related to combinatorial problems", *Linear Algebra and its Applications* 2 (1969) 451–558.

[6] R.E. Gomory, "Properties of a class of integer polyhedra", in: *Integer and nonlinear programming*, Ed. J. Abadie (North-Holland, Amsterdam, 1970) pp. 353–365.

[7] R.E. Gomory and E.L. Johnson, "Some continuous functions related to corner polyhedra, I", *Mathematical Programming* 3 (1972) 23–86.

[8] R.E. Gomory and E.L. Johnson, "Some continuous functions related to corner polyhedra, II", *Mathematical Programming* 3 (1972) 359–389.

[9] J.H. Griesmer, "A bound for error-correcting codes", *IBM Journal of Research and Development* 4 (1960) 532–542.

[10] H.J. Helgert and R.D. Stinoff, "Minimum-distance bounds for binary linear codes", *IEEE Transactions on Information Theory* 19 (1973) 344–356.

[11] W.W. Peterson and E.J. Weldon, Jr., *Error-correcting codes*, 2nd ed. (The MIT Press, Cambridge, Mass., 1972).

[12] D. Slepian, "A class of binary signalling alphabets", *Bell System Technical Journal* 35 (1956) 203–234.

[13] G. Solomon, "Algebraic coding theory", in: *Communication Theory*, Ed. A.V. Balakrishnan (McGraw-Hill, New York, 1968) ch. 5, pp. 216–271.

[14] G. Solomon and J. Stiffler, "Algebraically punctured codes", *Information and Control* 8 (1965) 170–179.

[15] A.W. Tucker, "A combinatorial equivalence of matrices", in: *Combinatorial analysis*, Proceedings of Symposia in Applied Mathematics X, Eds. R. Bellman and M. Hall (Am. Math. Soc., Providence, R.I., 1960) pp. 129–140.

[16] A.W. Tucker, "Combinatorial theory underlying linear programs", in: *Recent advances in mathematical programming*, Eds. R.L. Graves and P. Wolfe (McGraw-Hill, New York, 1963) pp. 1–16.

Mathematical Programming Study 1 (1974) 148–158. North-Holland Publishing Company

A NEW PROOF OF THE FUNDAMENTAL THEOREM OF ALGEBRA

H.W. KUHN*

Princeton University, Princeton, N.J., U.S.A.

Received 7 February 1974
Revised manuscript received 6 March 1974

Dedicated to Al Tucker, inspired teacher, valued colleague, and treasured friend, with the hope that he will see in it the results of his instruction and example

A constructive proof of the Fundamental Theorem of Algebra is given based on a labelling procedure and a combinatorial lemma applying to labelled triangulations of regions of the complex plane.

1. Introduction

This paper presents a new proof of the Fundamental Theorem of Algebra based on a labelling procedure and a combinatorial lemma applying to labelled triangulations of regions of the complex plane. Section 2 describes the rules for labelling, states the Main Lemma, and gives the derivation of the theorem from it. Section 3 is devoted to the proof of the Main Lemma. Section 4 contains observations on work related to the proof presented in this paper.

2. Labelling, roots, and the main lemma

Let $f(z)$ be a monic polynomial of degree n in the complex variable z with complex numbers as coefficients, that is, $f(z) = z^n + a_1 z^{n-1} +$

* This research was supported by the National Science Foundation under Grant GP-35698X.

$\ldots + a_{n-1}z + a_n$, where $a_1, \ldots, a_n$ are complex constants. Then the Fundamental Theorem of Algebra asserts that there exists at least one $\bar{z}$ with $f(\bar{z}) = 0$. Such a value $\bar{z}$ is called a *root* of $f(z)$.

The polynomial $f(z)$ induces a *labelling* of the z-plane. If $f(z) \neq 0$ and $f(z) = u + iv$, define $\arg f(z)$ to be the unique angle α such that $-\pi < \alpha \leq \pi$ and $\cos \alpha = u/\sqrt{u^2 + v^2}$, $\sin \alpha = v/\sqrt{u^2 + v^2}$. The $\arg f(z)$ determines the *label* of z, $\ell(z)$, by the following rules:

$$\ell(z) = \begin{cases} 1 & \text{if } -\pi/3 \leq \arg f(z) \leq \pi/3 \text{ or } f(z) = 0, \\ 2 & \text{if } \pi/3 < \arg f(z) \leq \pi, \\ 3 & \text{if } -\pi < \arg f(z) < -\pi/3. \end{cases}$$

Our first lemma establishes the connection between the labelling induced by $f(z)$ and the roots of $f(z)$.

Lemma 2.1. *If z_1, z_2, z_3 are such that $\ell(z_k) = k$ and $|f(z_j) - f(z_k)| \leq \varepsilon$ for $j, k = 1, 2, 3$, then $|f(z_k)| \leq 2\varepsilon/\sqrt{3}$.*

Proof. Fig. 1 is drawn in the w-plane of values of $f(z)$. Regions which induce $\ell(z) = 1, 2, 3$ are labelled accordingly. From the hypotheses of Lemma 2.1, the shaded set A exhibits all $w = f(z)$ with $\ell(z) = 1$ which are within ε of regions 2 and 3. Hence $|f(z_1)| \leq 2\varepsilon/\sqrt{3}$ and a similar argument can be made for $k = 2$ and 3.

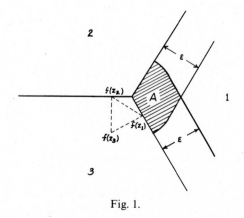

Fig. 1.

The three values $f(z_1), f(z_2), f(z_3)$ shown prove that the inequality of the lemma is the best possible.

If we restrict our search for a root to a fixed bounded set in the z-plane, the uniform continuity of $f(z)$ will ensure that small triangles $\{z_1, z_2, z_3\}$ in this set are mapped into small triangles $\{f(z_1), f(z_2), f(z_3)\}$ in the w-plane. If the small triangle $\{z_1, z_2, z_3\}$ carries a complete set of labels 1, 2, and 3, then Lemma 2.1 tells us that all of the $f(z_k)$ are small. Thus we are led naturally to hunt for very small triangles in the z-plane with a complete set of labels. To obtain a large supply of triangles in the plane, consider the square grid with side $h > 0$ and divide each square along a diagonal as shown in Fig. 2.

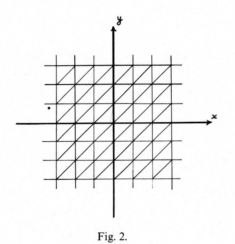

Fig. 2.

We shall also need a finite square, centered at the origin and triangulated in this way. Let $Q_{m,h}$ denote the region of the $z = x + iy$ plane, triangulated as above and such that $|x| \le R$ and $|y| \le R$ for $R = mh$. We are now in a position to outline the proof of the Fundamental Theorem of Algebra. With a fair amount of tedious analysis, we shall prove:

Lemma 2.2 (Main Lemma). *There exists a fixed R such that, for all m large enough, every $Q_{m,h}$ contains a completely labelled triangle.*

The Fundamental Theorem of Algebra follows directly from the Main Lemma. For m large enough, let $\{z_1(m), z_2(m), z_3(m)\}$ be a completely labelled triangle in $Q_{m,h}$. Since $Q_{m,h}$ is a compact set, the sequence

$z_1(m)$ contains a convergent subsequence $z_1(m_k)$ with limit $\bar{z}$. Since $mh = R$ is fixed, $h \to 0$ as $m \to \infty$, and Lemma 2.1 implies

$$f(\bar{z}) = \lim_{k \to \infty} f(z_1(m_k)) = 0.$$

Therefore, for the purposes of the existence proof, we can restrict our attention to the proof of the Main Lemma.

3. Proof of the Main Lemma

Let each of the triangles Δ in the triangulated square $Q_{m,h}$ be oriented with the usual positive orientation of the plane. This induces an orientation of the boundary $\partial Q_{m,h}$ of the square, which is subdivided into $8m$ directed edges (z_1, z_2) of length h. Each such edge carries an ordered pair of labels (ℓ_1, ℓ_2).

Lemma 3.1. *If there is at least one edge labelled* $(1, 2)$ *and no edge labelled* $(2, 1)$ *on* $\partial Q_{m,h}$, *then* $Q_{m,h}$ *contains a completely labelled triangle.*

Proof. Construct a sequence of *distinct* triangles $\Delta_1, \ldots, \Delta_s, \ldots$ *inside* $Q_{m,h}$ as follows. The first triangle, Δ_1, is obtained by taking an edge labelled $(1, 2)$ on $\partial Q_{m,h}$ and completing it with a third vertex to the unique triangle inside $Q_{m,h}$. Suppose distinct triangles $\Delta_1, \ldots, \Delta_s$ have been constructed inside $Q_{m,h}$, each with a directed edge labelled $(1, 2)$ on its (oriented) boundary. If the third vertex of Δ_s is labelled 3, then Δ_s is completely labelled and the construction ends. Otherwise, the third label duplicates exactly one of the other two labels. Drop the corresponding vertex and let Δ_{s+1} be the unique triangle adjacent to Δ_s and containing the two remaining vertices. The two cases shown in Fig. 3 prove

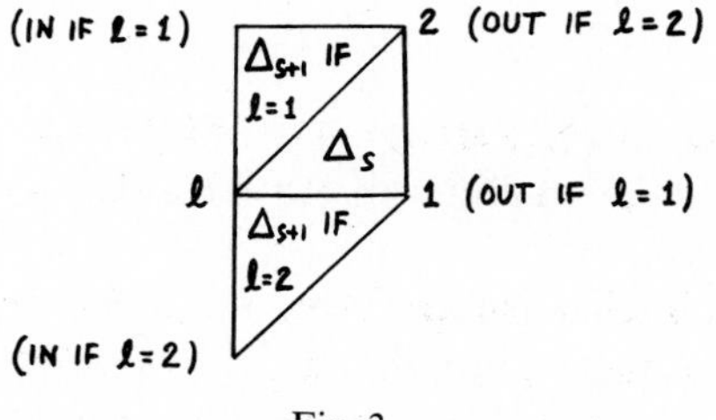

Fig. 3.

that these vertices span a directed edge labelled $(1, 2)$ on the (oriented) boundary of Δ_{s+1}.

We now must verify that Δ_{s+1} is inside $Q_{m,h}$ and distinct from Δ_1, ..., Δ_s. Since Δ_s is inside $Q_{m,h}$, if Δ_{s+1} were outside, then the common edge would carry the ordered pair of labels $(2, 1)$ in the orientation of the boundary of $Q_{m,h}$, contrary to assumption. If $\Delta_{s+1} = \Delta_1$, then Δ_s is outside $Q_{m,h}$, which is a contradiction. If $\Delta_{s+1} = \Delta_t$ for $2 \le t \le s$, then $\Delta_s = \Delta_{t-1}$, which is impossible since the previously constructed triangles were distinct. Finally, since there are only a finite number of triangles inside $Q_{m,h}$ the construction must terminate with a completely labelled triangle.

The constructive algorithm which underlies this proof is illustrated in Fig. 4.

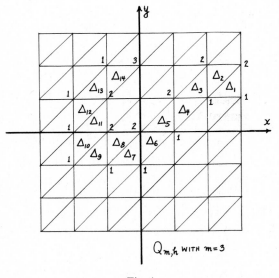

$Q_{m,h}$ WITH $m = 3$

Fig. 4.

As a consequence of this lemma, we may now restrict our attention to the properties of the labelling on the boundary of $Q_{m,h}$.

Lemma 3.2. *For R large enough,* $\ell(R) = 1$.

Proof. Rewrite $f(z) = z^n(1 + g(z))$, where $g(z) = a_1/z + \ldots + a_n/z^n$. Then

$$|g(R)| \le \frac{|a_1|}{R} + \ldots + \frac{|a_n|}{R^n} \le \frac{\max|a_k|}{R-1} \quad \text{for } R > 1.$$

Hence, if $R \ge 2\max|a_k|/\sqrt{3} + 1$, then $|g(R)| \le \frac{1}{2}\sqrt{3}$. Referring to Fig. 5, we have

$$|\arg f(R)| = |\arg(1 + g(R))| \le \frac{\pi}{3}$$

which implies $\ell(R) = 1$ for $R \ge 2\max|a_k|/\sqrt{3} + 1$.

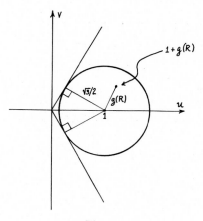

Fig. 5.

Lemma 3.3. *For a fixed R large enough,*

$$\frac{\pi}{12m} < \arg\frac{f(z_2)}{f(z_1)} < \frac{2\pi}{3}$$

for all $(z_1, z_2) \in \partial Q_{m,h}$ and all m sufficiently large.

Proof. Factoring $f(z) = z^n(1 + g(z))$, we shall work on

$$\arg\frac{z_2^n}{z_1^n} \quad \text{and} \quad \arg\frac{1 + g(z_2)}{1 + g(z_1)}$$

separately. Let $(z_1, z_2) \in \partial Q_{m,h}$ and $\theta = \arg z_2/z_1$. Then it is clear from Fig. 6 that $\alpha \le \theta \le \beta$, where

$$\sin \alpha = \frac{\frac{1}{2}\sqrt{2}}{\sqrt{m^2 + (m-1)^2}} \quad \text{and} \quad \tan \beta = \frac{1}{m}.$$

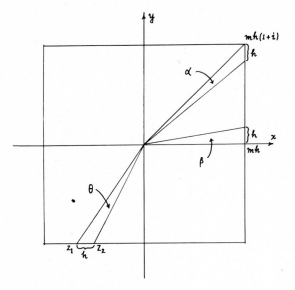

Fig. 6.

Hence

$$1/2m < \alpha \le \theta \le \beta < 1/m$$

and

$$n/2m < \arg z_2^n / z_1^n < n/m \quad \text{for } m \ge n/\pi. \tag{1}$$

(Note that the restriction $m \ge n/\pi$ is a consequence of the choice $-\pi < \arg w \le \pi$.)

To obtain similar bounds on $\arg (1 + g(z_2))/(1 + g(z_1))$, first note that

$$\frac{1 + g(z_2)}{1 + g(z_1)} = 1 + \frac{g(z_2) - g(z_1)}{1 + g(z_1)} \tag{2}$$

and

$$|g(z)| \le \frac{|a_1|}{R} + \ldots + \frac{|a_n|}{R^n} \le \frac{\max |a_k|}{R - 1} \quad \text{for } |z| \ge R > 1,$$

while

$$|g(z_2) - g(z_1)| \leq h\left(\frac{|a_1|}{R^2} + \frac{2|a_2|}{R^3} + \ldots + \frac{n|a_n|}{R^{n+1}}\right)$$

$$\leq \frac{h \max |a_k|}{(R-1)^2} \quad \text{for } (z_1, z_2) \in \partial Q_{m,h}, \quad R > 1.$$

Noting that $R \geq 4$ implies $1/(R-1)^2 \leq 2/R^2$ and $R \geq 2 \max |a_k| + 1$ implies $|g(z_1)| \leq \frac{1}{2}$ we have

$$\left|\frac{g(z_2) - g(z_1)}{1 + g(z_1)}\right| \leq \frac{|g(z_2) - g(z_1)|}{1 - |g(z_1)|} \leq \frac{4 \max |a_k|}{mR} \leq 1 \tag{3}$$

for $R \geq 4 \max |a_k| + 4$ and all $m \geq 1$. Referring to Fig. 7, $|w| \leq 1$ implies $-\pi/2 \leq \arg(1 + w) \leq \pi/2$ and $\sin \arg(1 + w) \leq |w|$.

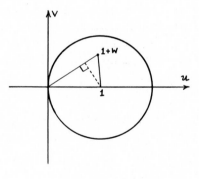

Fig. 7.

Hence $|w| \leq 1$ implies $|\arg(1 + w)| \leq 2|w|$ and, by (2) and (3),

$$\left|\arg\left(\frac{1 + g(z_2)}{1 + g(z_1)}\right)\right| \leq \frac{8 \max |a_k|}{mR} \tag{4}$$

for $R \geq 4 \max |a_k| + 4$ and all $m \geq 1$.

Combining (1) and (4), we have

$$\frac{1}{m}\left(\frac{n}{2} - \frac{8 \max |a_k|}{R}\right) \leq \arg \frac{f(z_2)}{f(z_1)} \leq \frac{1}{m}\left(n + \frac{8 \max |a_k|}{R}\right)$$

for all $(z_1, z_2) \in \partial Q_{m,h}$ provided that $R \geq 4 \max |a_k| + 4$, $m \geq n/\pi$, and that the lower bound does not exceed π. If we choose $m \geq \frac{1}{2}(n + 2)$, then $m > 3/2\pi \, (n + 8 \max |a_k|/R)$ and hence

$$\arg \frac{f(z_2)}{f(z_1)} < \frac{2\pi}{3}. \tag{5}$$

If we set $R = 48 \max |a_k| + 4$, then $R > 8 \max |a_k|/(n/2 - \pi/12)$ for all $n \geq 1$, which implies

$$\arg \frac{f(z_2)}{f(z_1)} \geq \frac{1}{m} \left(\frac{n}{2} - \frac{8 \max |a_k|}{R} \right) > \frac{\pi}{12m}$$

and completes the proof of the lemma.

It only remains to put the pieces together in the correct order. Choose $R = 48 \max |a_k| + 4$ and $m \geq \frac{1}{2}(n + 2)$. Lemma 3.2 asserts that $\ell(R) = 1$. Lemma 3.3 asserts that, as we traverse the $8m$ directed edges on the boundary of $Q_{m,h}$, successive labels never reverse their proper cyclic order (because of (5)) and there must be at least one change of label from 1 to 2, since

$$\arg \frac{f(z_2)}{f(z_1)} > \frac{\pi}{12m} \quad \text{and} \quad 8m \frac{\pi}{12m} = \frac{2\pi}{3}.$$

This gives the hypotheses of Lemma 3.1 and assures the existence of a completely labelled triangle in $Q_{m,h}$ for R fixed and all m sufficiently large. This is exactly the assertion of the Main Lemma.

4. Observations

The idea that a labelling procedure combined with the appropriate combinatorial lemma could be used to prove the Fundamental Theorem of Algebra occurred to me 15 years ago in connection with work on Sperner's Lemma and the Brouwer Fixed Point Theorem [3]. The difficulties of establishing a direct connection are well known (see, for example, the invalid derivation of the Fundamental Theorem of Algebra from Brouwer's Theorem in [1]). The conviction that such an approach must work was strengthened by the superb example set by Tucker's paper [11] which presents an elegant combinatorial lemma applied to existence proofs from topology.

The stimulus to return to the problem came from the recent success of constructive combinatorial algorithms to approximate fixed points (see [10, and 4]). The construction used in the proof of Lemma 3.1 is an oriented version of the construction used in [4] and the proof can be cast in the language of oriented Sperner graphs used in [5]. Namely, a triangle Δ with 1 and 2 among its labels is called a *node*. A *directed edge* joins Δ_1 to an adjacent Δ_2 if their common side is labelled (2, 1) on the oriented boundary of Δ_1 and (1, 2) on the oriented boundary of Δ_2. *Sources* are triangles with a directed edge (1, 2) on $\partial Q_{m,h}$ or triangles with labels (1, 3, 2) in cyclic order. *Sinks* are triangles with a directed edge (2, 1) on $\partial Q_{m,h}$ or triangles with labels (1, 2, 3) in cyclic order. Then the construction of Lemma 3.1 (paralleling the Algorithm of [5, p. 202]) leads us from a source with an edge on $\partial Q_{m,h}$ to a sink with a complete set of labels with the cyclic order (1, 2, 3). (Of course, the argument that no triangle is examined twice echos the arguments used for the convergence of the Simplex Method or used the work of Lemke and Howson in [7] and [8].) It should be noted that this formulation leads directly to the following "Combinatorial Stokes' Theorem": Let Q be a bounded region of the plane (connected or not) subdivided into triangles with a positive orientation. Suppose that the vertices of the subdivision are labelled 1, 2 or 3. Define $\sigma(\Delta) = +1$ or -1 if $\ell(\Delta) = (1, 2, 3)$ or (1, 3, 2), and $\sigma(\Delta) = 0$ otherwise. For a directed edge $e \subset \partial Q$, let $\sigma(e) = +1$ or -1 if $\ell(e) = (1, 2)$ or (2, 1), and $\sigma(e) = 0$ otherwise. Then

$$\sum_{\Delta \subset Q} \sigma(\Delta) = \sum_{e \subset \partial Q} \sigma(e).$$

Although the purpose of this paper is strictly to establish existence, the constructive nature of the proof suggests the computation of roots by the same procedure. An experimental computer program incorporating these ideas was written in May 1973 by E. Rosenthal and has found complete sets of roots for a variety of test polynomials. The program finds *all* roots by starting the algorithm of Lemma 3.1 at each of the n edges labelled (1, 2) on $\partial Q_{m,h}$ (constructed in such a manner to insure the existence of these edges and no reversals (2, 1)). When a triangle labelled (1, 2, 3) is found, the computer code refines the subdivision using techniques similar to those employed by Eaves [9] or Kuhn and Mac-Kinnon [6]. A recent improvement in the algorithm eliminates the use of theoretical bounds to choose R for a given polynomial and starts near the origin rather than on the boundary of the square. The revised algorithm accepts as input the coefficients $a_1, \ldots, a_n$ and produces as output n sequences converging to the n roots of the polynomial. Except for trivial bookkeeping, the generation of a new point in any sequence requires one evaluation of the polynomial. (Compare the constructive proof of the Fundamental Theorem of Algebra given by Rosenbloom [9].) Details of these algorithms and computer programs will be published elsewhere.

Added in proof

The elegant paper of Ky Fan [6a] includes the Combinatorial Stokes, Theorem of Section 4 as a special case of Corollary 2 (p. 600).

References

[1] B.H. Arnold, "A topological proof of the fundamental theorem of algebra", *American Mathematical Monthly* 56 (1949) 465–466.

[2] B.C. Eaves, "Homotopies for the computation of fixed points", *Mathematical Programming* 3 (1972) 1–23.

[3] H.W. Kuhn, "Some combinatorial lemmas in topology", *IBM Journal of Research and Development*, 4 (1960) 518–24.

[4] H.W. Kuhn, "Simplicial approximation of fixed points", *Proceedings of the National Academy of Sciences* 61 (1968) 1238–42.

[5] H.W. Kuhn, "Approximate search for fixed points", in: *Computing methods in optimization problems* 2, Eds. L.A. Zadeh, L.W. Neustadt, A.V. Balakrishnan (Academic Press, New York, 1969) pp. 199–211.

[6] H.W. Kuhn and J.G. MacKinnon, "The Sandwich Method for finding fixed points", to appear.

[6a] Ky Fan, "Simplicial maps from an orientable n-pseudomanifold into S^m with the octahedral triangulation", *Journal of Combinatorial Theory* 2 (1967) 588–602.

[7] C.E. Lemke, "Bimatrix equilibrium points and mathematical programming", *Management Science* 11 (1964-5) 681–9.

[8] C.E. Lemke and J.T. Howson, Jr., "Equilibrium points of bimatrix games", *Journal of the Society for Industrial and Applied Mathematics* 12 (1964) 412–23.

[9] P.C. Rosenbloom, "An elementary constructive proof of the fundamental theorem of algebra", *American Mathematical Monthly* 52 (1945) 562–570.

[10] H. Scarf, "The approximation of fixed points of a continuous mapping", *SIAM Journal of Applied Mathematics* 15 (1967) 1328–43.

[11] A.W. Tucker, "Some topological properties of disk and sphere", in: *Proceedings of the first Canadian mathematical congress*, Montreal, 1945 (The University of Toronto Press, Toronto, 1946) pp. 285–309.

Mathematical Programming Study 1 (1974) 159–174. North-Holland Publishing Company

PIVOTAL THEORY OF DETERMINANTS

Stephen B. MAURER

University of Waterloo, Waterloo, Ontario, Canada

Received 17 January 1974

In honor of Albert W. Tucker

With the aid of Tucker's pivotal algebra we show that determinants may be defined in a way which is both computationally efficient and theoretically manageable.

1. Introduction

Let $M = [a_{ij}]$ be an $n \times n$ matrix over field F. There are many ways to define the determinant $|M|$. First, there is the sum over the symmetric group:

$$\sum_{\sigma \in S_n} (-1)^\sigma \prod_{i=1}^{n} a_{i\,\sigma(i)}.$$

Second, there is an inductive definition using expansion by minors:

$$|M| = \sum_{i=1}^{n} (-1)^{i+j} a_{ij} |M_{ij}|, \tag{1}$$

where M_{ij} is M with row i and column j deleted. Third, there is the definition in terms of alternating forms, or equivalently, n-dimensional signed volume or n-fold products in a Grassmann algebra. While each of these definitions has its advantages, they all share one glaring fault: if one must actually compute a determinant they are terribly inefficient.

The first two methods require of the order of $n!$ computations, and the third does not immediately offer any method of computation at all.

One of Professor Tucker's strong convictions is that a good definition of a computable object is one with which the object may be computed directly and efficiently, yet which is also theoretically manageable. Our purpose here is to provide such a definition for determinants, and give examples of its use.

Our definition is hardly new; an efficient *procedure* for evaluating determinants is well known, and this, suitably stated, is what we will use. What is new, we believe, is its use as a definition.

Let us recall this well-known procedure. Given an $n \times n$ matrix M, pick some particular entry $a_{ij} \neq 0$ and set $f_1 = (-1)^{i+j} a_{ij}$. Next, make all the other entries of column j zero by subtracting multiples of row i from the other rows. Then delete row i and column j to obtain an $(n-1) \times (n-1)$ matrix N. Repeat the process on N, and so on, until obtaining a matrix of zeros or until no matrix is left. In the former case $|M| = 0$, in the latter $|M|$ is the product of the f's. (In practice, one terminates at a matrix with even one row or column of zeros, or at a 2×2 matrix.) This method is called *pivotal condensation*. It was introduced by Chio in 1853, but was implicit in earlier work of Gauss. It requires about $n^3/3$ multiplications.

(2) provides an example, which we will soon compare with other pivotal procedures applied to the same initial matrix. The special elements used to determine the f's are marked by asterisks and are called the *pivots*.

$$\begin{bmatrix} 6 & 2 & 3 \\ 2^* & 1 & 2 \\ 4 & 0 & 1 \end{bmatrix} \longrightarrow \begin{bmatrix} -1^* & -3 \\ -2 & -3 \end{bmatrix} \longrightarrow [3] \longrightarrow \emptyset, \qquad (2)$$

$$f_1 = -2, \qquad f_2 = -1, \qquad f_3 = 3, \qquad |M| = 6.$$

The reason pivotal condensation has not been used as a definition should be clear. There are so many orders in which one can do a condensation of M; how does one know they all give the same value? They do, of course. Given those approaches which define determinants uniquely to begin with, the constant value of condensations follows from elementary properties of determinants that are then easy to prove, namely (1) and the row analog of 4.2. However, these properties do not even make sense until one has a unique determinant to refer to, so if we are going to use condensation for our definition, we will have to prove uniqueness from scratch. This is the main enterprise of our paper.

Note. The use of (1) as a definition also suffers a uniqueness problem, but that problem can be circumvented by initially fixing j, say $j = 1$. See Noble [3, Section 7.2]. For condensation such a trick fails. We might initially insist on pivoting always, say, at the top left, but any such predetermined strategy may be blocked by a zero pivot even though $|M| \neq 0$.

2. Pivot sequences and pivotal algebra

To prove uniqueness, we will embed pivotal condensation in the inversion procedure of Tucker's pivotal algebra. To motivate this procedure, we first recall two better known procedures, Gaussian elimination and Gauss–Jordan elimination.

Given M, not necessarily square, Gaussian elimination proceeds as follows. Pick some row i and column j with non-zero common entry and reduce all the other entries of column j to 0 by subtracting multiples of row i from the others. Then pick some other row i' and column j' whose common entry is now non-zero and reduce to 0 all entries of column j' not in rows i, i' by subtracting multiples of row i'. Continue until there are no more unpicked rows, or no more unpicked columns, or no more non-zero entries common to the unpicked rows and columns. For example,

$$\begin{bmatrix} 6 & 2 & 3 \\ 2^* & 1 & 2 \\ 4 & 0 & 1 \end{bmatrix} \longrightarrow \begin{bmatrix} 0 & -1^* & -3 \\ 2 & 1 & 2 \\ 0 & -2 & -3 \end{bmatrix} \longrightarrow \begin{bmatrix} 0 & -1 & -3 \\ 2 & 1 & 2 \\ 0 & 0 & 3^* \end{bmatrix} \longrightarrow \text{no change.} \quad (3)$$

Clearly, pivotal condensation is exactly the same thing, except that at each step attention is restricted to the intersection of the so-far un-picked rows and columns. Thus the matrices of (2) are submatrices, respectively, of the matrices of (3).

Gauss–Jordan elimination includes all the operations of Gaussian elimination, but also that at each step the pivot element is reduced to 1 by multiplying the pivot row by a constant, and *all* other entries in the pivot column are reduced to 0. So instead of (3) we get

$$\begin{bmatrix} 6 & 2 & 3 \\ 2^* & 1 & 2 \\ 4 & 0 & 1 \end{bmatrix} \longrightarrow \begin{bmatrix} 0 & -1^* & -3 \\ 1 & \frac{1}{2} & 1 \\ 0 & -2 & -3 \end{bmatrix} \longrightarrow \begin{bmatrix} 0 & 1 & 3 \\ 1 & 0 & -\frac{1}{2} \\ 0 & 0 & 3^* \end{bmatrix} \longrightarrow \begin{bmatrix} 0 & 1 & 0 \\ 1 & 0 & 0 \\ 0 & 0 & 1 \end{bmatrix} \quad (4)$$

More formally, if $[b_{ij}]$ is obtained from $[a_{ij}]$ by a Gauss–Jordan pivot on $a_{k\ell}$, then

$$b_{k\ell} = 1, \qquad b_{i\ell} = 0, \qquad i \neq k; \tag{5}$$

$$\begin{aligned} b_{kj} &= a_{kj}/a_{k\ell}, \qquad j \neq \ell, \\ b_{ij} &= a_{ij} - (a_{i\ell}\,a_{kj}/a_{k\ell}), \qquad i \neq k, \quad j \neq \ell. \end{aligned} \tag{6}$$

Note that the matrices of (2) are also submatrices of those of (4).

Although Gauss–Jordan elimination involves more work than Gaussian elimination, it has advantages. First, its end result is sparser. More important, it has a simple property which has come to light since the advent of linear programming. If $[a_{ij}]$ is $m \times n$, associate with its rows some set $X = \{x_1, x_2, \ldots, x_m\}$ of linearly independent vectors (e.g., m independent variables). Now each column of M gives the co-ordinates of a vector y (e.g. linear dependent variable) in the span of X. We write

$$
\begin{array}{c}
x_1 \\ \vdots \\ x_k \\ \vdots \\ x_m
\end{array}
\left[
\begin{array}{ccc}
& \vdots & \\
\cdots & a_{k\ell} & \cdots \\
& \vdots &
\end{array}
\right]
\qquad (7)
$$
$$= y_1 \quad = y_\ell \quad = y_n$$

and call this a *bordered matrix* or *tableau*. The simple property is: if $[b_{ij}]$ is obtained from $[a_{ij}]$ by a Gauss–Jordan pivot on $a_{k\ell}$ then $[b_{ij}]$ is the unique matrix which can be bordered by replacing x_k by y_ℓ, that is,

$$
\begin{array}{c}
x_1 \\ \vdots \\ y_\ell \\ \vdots \\ x_m
\end{array}
\left[
\begin{array}{ccc}
& \vdots & \\
\cdots & b_{k\ell} & \cdots \\
& \vdots &
\end{array}
\right].
$$
$$= y_1 \quad = y_\ell \quad = y_n$$

To put it another way, if matrices are viewed as collections of column vectors, Gauss–Jordan elimination amounts to a change of basis one vector at a time until the y's are represented solely in terms of an independent subset of themselves.

The proof that Gauss–Jordan elimination has this bordered matrix interpretation is simple linear algebra: solve $y_\ell = \sum_{i=1}^{m} a_{i\ell} x_i$ for x_k and substitute this solution in the equations for the other y's.

When viewed in terms of bordered matrices, Gauss–Jordan elimination is asymmetric. It would be nice to have a third pivot procedure that interchanges x_k and y_ℓ. Clearly the new rules would include (6). A little more linear algebra takes care of column ℓ too:

$$\begin{aligned} b_{k\ell} &= 1/a_{k\ell} \\ b_{i\ell} &= -a_{i\ell}/a_{k\ell}, \qquad i \neq k. \end{aligned} \tag{8}$$

An easy way to remember (6) and (8) is

$$\boxed{\begin{array}{cc} p^* & q \\ r & s \end{array}} \longrightarrow \boxed{\begin{array}{cc} 1/p & q/p \\ -r/p & s - (q\,r/p) \end{array}} \tag{9}$$

where the second row (column) represents any row (column) not containing the pivot.

Here is the analog to (2); clearly, (2) is embedded in it.

$$
\begin{array}{c|ccc}
x_1 & 6 & 2 & 3 \\
x_2 & 2^* & 1 & 2 \\
x_3 & 4 & 0 & 1 \\
\hline
 & = y_1 & = y_2 & = y_3
\end{array}
\longrightarrow
\begin{array}{c|ccc}
x_1 & -3 & -1^* & -3 \\
y_1 & \frac{1}{2} & \frac{1}{2} & 1 \\
x_3 & -2 & -2 & 3 \\
\hline
 & = x_2 & = y_2 & = y_3
\end{array}
\longrightarrow
$$

$$
\begin{array}{c|ccc}
y_2 & 3 & -1 & 3 \\
y_1 & -1 & \frac{1}{2} & -\frac{1}{2} \\
x_3 & 4 & -2 & 3^* \\
\hline
 & = x_2 & = x_1 & = y_3
\end{array}
\longrightarrow
\begin{array}{c|ccc}
y_2 & -1 & 1 & -1 \\
y_1 & -\frac{1}{3} & \frac{1}{6} & \frac{1}{6} \\
y_3 & \frac{4}{3} & -\frac{2}{3} & \frac{1}{3} \\
\hline
 & = x_2 & = x_1 & = x_3
\end{array}\,.
$$

$$\tag{10}$$

The use of this third type of pivot step has been named by Tucker *pivotal algebra*. By emphasizing the border vectors, he has made this algebra a simple but powerful tool [5]. These vectors allow one to get results about the pivot procedure quickly by applying standard theorems of linear algebra, or conversely, to obtain constructive proofs of these theorems by using the pivot procedure.

From now on, all our pivot steps will be of this third type.

Let us observe several facts about pivotal algebra of special use to us. First, given $M = [a_{ij}]$ as in (7), with $X = \{x_1, \ldots, x_m\}$ and $Y = \{y_1, \ldots, y_n\}$, then X is clearly a basis of $V = \text{span}(X \cup Y)$. Moreover, $a_{k\ell} \neq 0$, that is, $a_{k\ell}$ can be used as a pivot if and only if $X - x_k + y_\ell$ is also a basis of V. More generally, if M' can be obtained from M by a finite sequence of pivots and $Z \subset X \cup Y$ is the set of vectors indexing rows of M', then Z is a basis of V. Such an M' is said to be *combivalent* to M. Now suppose M', M'' are both combivalent to M and both have row border Z. Since the columns of both express the vectors $W = (X \cup Y) - Z$ as linear combinations over Z, $M' = M''$, except that in as much as the orders of Z and W as borders on M' and M'' may differ, the orders of the entries of M' and M'' may differ correspondingly. However, such differences in order can only involve vectors actually exchanged in the pivot sequences leading from M to M' and M''; thus the submatrices of M', M'' indexed by vectors not pivoted in either sequence are *exactly* the same.

Now suppose M is $n \times n$ and, as in (10), X and Y can be completely exchanged to obtain M'. Since M expresses Y in terms of X and M' expresses X in terms of Y, up to order $M' = M^{-1}$. Henceforth a sequence of n pivots leading from M to M^{-1} will be called an *inversion sequence*.

Finally, there is a converse to a statement made above. If $Z \subset X \cup Y$ is a basis of V, then there exists M' combivalent to M with row border Z. This follows by repeated use of the following well-known and easily proved

Axiom 2.1 (Exchange Axiom). *If B, B' are bases of vector space V and $b \in B - B'$, there exists $b' \in B' - B$ so that $B - b + b'$ is also a basis of V.*

Let us summarize the import of pivotal algebra for our determinant problem. There is a one-to-one embedding of complete pivotal condensations of M (that is, those not terminated by a zero matrix) into inversion sequences for M. Thus at least one complete condensation exists if and only if M^{-1} exists. Henceforth, for theoretical purposes we will deal only with inversion sequences.

Note. There is one other ingredient in pivotal algebra which we ought to mention, though we won't use it. Border vectors can also be put on top and to the right side of a matrix. If each one on the right represents the *negative* of its row as a linear combination of the top vectors, then these too get exchanged in pairs by pivot steps. In linear programming this allows one to represent and solve primal and dual programs simultaneously.

3. Proof of uniqueness

It is time for some formality!

Definition 3.1. Let M be an $n \times n$ matrix and σ, τ permutations of $\{1, 2, \ldots, n\}$. Then (M, σ, τ) denotes the inversion sequence $M_1 = M$, $M_2, \ldots, M_{n+1} = M^{-1}$, where M_{k+1} is obtained from M_k by pivoting on the (σ_k, τ_k) entry of the latter. This pivot entry is denoted $p_k(M, \sigma, \tau)$.

For a given σ, τ, (M, σ, τ) may not exist, even if M^{-1} does. Naturally, all statements to follow are about those that do.

$p_k(M, \sigma, \tau)$ is also an entry of the kth condensation matrix embedded in (M, σ, τ), and we need names for the indices of p_k in this smaller matrix. Let us call them (s_k, t_k). Clearly

$$s_k = \sigma_k - \text{card} \{\sigma_i \colon i < k, \sigma_i < \sigma_k\}, \tag{11}$$

and a similar statement holds for t_k.

Definition 3.2. $|M, \sigma, \tau|$, the *determinant* of (M, σ, τ), is

$$\prod_{k=1}^{n} (-1)^{s_k + t_k} p_k(M, \sigma, \tau).$$

It should be clear that this is a formalization of the method of pivotal condensation. We must prove

Theorem 3.3. *For any (M, σ, τ) and (M, σ', τ'), $|M, \sigma, \tau| = |M, \sigma', \tau'|$.*

The idea of our proof is simple. First we will define a notion of elementary deformation of inversion sequences. Then we will show (i) two inversions differing by an elementary deformation have the same determinant (Theorem 3.5), and (ii) any two inversions whatever differ by a finite chain of elementary deformations (Theorem 3.6).

Definition 3.4. (M, σ, τ) and (M, σ', τ') *differ by an elementary deforma-tion if* $\sigma = \sigma'$, *and* $\tau = \tau'$ *except possibly at two integers* k *and* $k + 1$. We write $(M, \sigma, \tau) \sim (M, \sigma', \tau')$.

Theorem 3.5. *If* $(M, \sigma, \tau) \sim (M, \sigma', \tau')$, *then* $| M, \sigma, \tau| = | M, \sigma', \tau' |$.

Proof. Suppose (M, σ, τ) and (M, σ', τ') differ on k, $k + 1$. Writing p_i, p'_i for $p_i(M, \sigma, \tau)$, $p_i(M, \sigma', \tau')$, it is clear that $p_i = p'_i$ when $i < k$, for the first $k - 1$ matrices in both sequences are pair by pair the same. $p_i = p'_i$ also holds when $i \geq k + 2$, for, as explained earlier (using border vectors), $M_{k+2}, \ldots, M_{n+1}$ in both sequences are pair by pair the same up to order of previously pivoted rows and columns. Moreover, from (11) it is easy to see that $s_i = s'_i$ and $t_i = t'_i$ for $i < k$ and $i \geq k + 2$. Thus we need only show

$$\prod_{i=k}^{k+1} (-1)^{s_i + t_i} p_i = \prod_{i=k}^{k+1} (-1)^{s'_i + t'_i} p'_i. \tag{12}$$

Let us rename $\sigma_k = i$, $\sigma_{k+1} = i'$, $\tau_k = j$, $\tau_{k+1} = j'$. Then steps k, $k + 1$ of (σ', τ') amount to one of three cases:
    (1) pivot on (i, j'), then on (i', j),
    (2) (i', j), then (i, j'),
    (3) (i', j'), then (i, j).

Order of rows and columns does not affect p_i and p'_i, so let us write the crucial 2×2 submatrix of the common M_k as

$$
\begin{array}{c c}
i & \boxed{\begin{array}{cc} p & q \\ r & s \end{array}} \\[2pt]
i' & \\[-2pt]
& \ \ j \ \ j'
\end{array}
$$

Then by direct calculation from (9) we get

$$p_k\, p_{k+1} = p\,s - q\,r, \tag{13}$$

$$p'_k\, p'_{k+1} = \begin{cases} q\,r - p\,s & \text{in cases (1) and (2),} \\ p\,s - q\,r & \text{in case (3).} \end{cases} \tag{14}$$

Now consider the powers of -1. If $\sigma = \sigma'$ (case 1), $s = s'$. If $\sigma \neq \sigma'$, either $i < i'$ so by (11) $s'_k = s_{k+1} + 1$, $s'_{k+1} = s_k$, or else $i > i'$ so $s'_k = s_{k+1}$, $s'_{k+1} = s_k - 1$. Either way we get

$$(-1)^{s_k + s_{k+1}} = -(-1)^{s'_k + s'_{k+1}}. \tag{15}$$

Similarly,

$$(-1)^{t_k + t_{k+1}} = -(-1)^{t'_k + t'_{k+1}} \tag{16}$$

just in cases (1) and (3). Thus there is a sign difference between (13) and (14) just when exactly one of (15) and (16) holds. This proves (12).

Theorem 3.6. *For any* (M, σ, τ), (M, σ', τ'), *there exist inversions* (M, σ_k, τ_k), $k = 1, 2, \ldots, p$, *such that* $(\sigma_1, \tau_1) = (\sigma, \tau)$, $(\sigma_p, \tau_p) = (\sigma', \tau')$, *and* (M, σ_k, τ_k) $\sim (M, \sigma_{k+1}, \tau_{k+1})$.

Proof. Suppose M has its rows bordered by X and its columns by Y. Then there is a one-to-one correspondence between inversion sequences of M and sequences

$$X = B_1 \xrightarrow{x_1, y_1} B_2 \ldots B_n \xrightarrow{x_n, y_n} B_{n+1} = Y \tag{17}$$

of bases of span $(X \cup Y)$, where $x_1, \ldots, x_n$ $(y_1, \ldots, y_n)$ is some ordering of X (Y) and

$$B \xrightarrow{x, y} B'$$

means $B' = B - x + y$. Namely, B_k is just the set of vectors indexing the rows of M_k. Moreover, (M, σ, τ) and (M, σ', τ') differ only at $k, k + 1$ if and only if their respective basis sequences agree, except for B_{k+1}. Consequently, given any two basis sequences (17) and

$$X \xrightarrow{x'_1, y'_1} B'_2 \ldots B'_n \xrightarrow{x'_n, y'_n} Y, \tag{18}$$

we need to demonstrate the existence of intermediate sequences each differing from the preceding one at just one basis. To do this we need the following lemma.

Lemma 3.7. *Given bases*

$$P \xrightarrow{x,\,y} Q \xrightarrow{x',\,y'} R,$$

with x, x', y, y' all distinct, there exists bases S, T, not necessarily distinct, such that

$$
\begin{array}{ccc}
 & P \xrightarrow{\ x',\,w\ } S & \\
(z,\,y') \big\downarrow & & \big\downarrow (x,\,w') \qquad\qquad (19)\\
 & T \xrightarrow{\ z',\,y\ } R &
\end{array}
$$

where $\{z, z'\} = \{x, x'\}$ and $\{w, w'\} = \{y, y'\}$.

Proof.   Applying the Exchange Axiom to P, R, there exists $S = P - x' + w$, where $w \in \{y, y'\}$; clearly, then $S - x + w' = R$, so S satisfies (19). Similarly, applying the axiom to R, P, there exists $T = R - y + z'$, where $z' \in \{x, x'\}$. Clearly, T satisfies (19). This proves the lemma.

   The significance of Lemma 3.7 is this: given any inversion sequence for which P, Q, R are consecutive bases, there is an elementary deformation (involving S) in which x' is pivoted in one step sooner, and an elementary deformation (involving T) in which y' is pivoted out one step sooner.
   Now, given any sequences $\mathscr{S}$, $\mathscr{S}'$ as in (17), (18), it suffices to show that if these first differ at pivot step k, then $\mathscr{S}'$ can be changed by finitely many elementary deformations into a third sequence which first differs from $\mathscr{S}$ at step $k + 1$. Without loss of generality we may assume $k = 1$.
   Clearly $x_1 = x_i'$ for some i. If $i > 1$, by the lemma there is an elementary deformation of $\mathscr{S}'$ in which x_1 is pivoted at step $i - 1$ instead. Using the lemma this way $i - 1$ times altogether, we get

$$\overline{\mathscr{S}}:\quad X \xrightarrow{x_1,\,\bar{y}_1} \bar{B}_2 \xrightarrow{\bar{x}_2,\,\bar{y}_2} \bar{B}_3 \ \ldots\ Y.$$

Now $y_1 = \bar{y}_j$ for some j. If $j = 1$, we are done. If $j = 2$, then $\bar{B}_3 = B_2 - \bar{x}_2 + \bar{y}_1$, so

$$X \xrightarrow{x_1, y_1} B_2 \xrightarrow{\bar{x}_2, \bar{y}_1} B_3 \xrightarrow{\bar{x}_3, \bar{y}_3} B_4 \ldots Y$$

is an elementary deformation of $\overline{\mathscr{S}}$ and we are done. Finally, if $j > 2$, by $j - 2$ applications of the lemma we may shift y_1 forward to the second step and then apply the previous case.

Note. In the last sentence we could have moved y_1 to the first step directly by $j - 1$ applications of the lemma, but in the last application we might also have dislodged x_1 from the first step.

We now know that the following is consistent:

Definition 3.8. If M is square, its determinant $|M|$ is

$$\begin{cases} |M, \sigma, \tau| & \text{if } (M, \sigma, \tau) \text{ exists}, \\ 0 & \text{if no inversion sequence for } M \text{ exists.} \end{cases}$$

However, in one respect we have yet to meet Tucker's criterion that the object defined should be directly computable from the definition, for this definition does not include an efficient way to tell when no (M, σ, τ) exists. Conceivably, one might try 100 different pairs (σ, τ), and be blocked by zeros each time, and yet simply be unlucky. Of course, we know this cannot happen, and it is easy to prove it using the border vectors. For suppose we are blocked by zeros on our *first* attempt, that is, after so many steps we get (up to order)

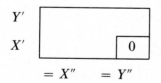

Indeed, suppose even one column of the (X', Y'') submatrix is zero. Then $Y = Y' \cup Y''$ is a dependent set, so M^{-1} does not exist, so no (M, σ, τ) exists. Thus we prefer

Definition 3.9. The determinant $|M|$ of the square matrix M is obtained as follows. Attempt to compute an inversion sequence of M. If this is successful, $|M| = |M, \sigma, \tau|$. If it is not, $|M| = 0$.

4. Pivotal proofs of standard properties

Theorem 4.1. *If N is obtained from M by multiplying column j by c, then* $|N| = c|M|$.

Proof. Here and in the next three theorems, either both M and N are invertible or neither is, and in the latter case the theorem is trivial. When M is invertible, row j cannot be all zeros, so we may pick (M, σ, τ) with $\tau_1 = j$. If the pivot from M to M_2 is schematized by (9), then the same pivot on N gives

$$
\begin{array}{|cc|}
\hline
c\,p^* & q \\[4pt]
c\,r & s \\
\hline
\end{array}
\quad \longrightarrow \quad
\begin{array}{|cc|}
\hline
1/(c\,p) & q/(c\,p) \\[4pt]
-r/p & s - q\,r/p \\
\hline
\end{array}
\;.
$$

The result is the same as M_2, except in the row and column of p. Consequently (N, σ, τ) exists too, and $p_k(N, \sigma, \tau) = p_k(M, \sigma, \tau)$ for $k \geq 2$. Finally, $p_1(N, \sigma, \tau) = c\,p = c\,p_1(M, \sigma, \tau)$.

Theorem 4.2. *Suppose N is obtained from M by adding c times column j to j'. Then* $|N| = |M|$.

Proof. Consider any (M, σ, τ) with $\tau_1 = j$. Letting (9) portray columns j, j' of M, M_2, then the same pivot on N gives

$$
\begin{array}{|cc|}
\hline
p^* & q + c\,p \\[4pt]
r & s + c\,r \\
\hline
\end{array}
\quad \longrightarrow \quad
\begin{array}{|cc|}
\hline
1/p & q/p + c \\[4pt]
-r/p & s - q\,r/p \\
\hline
\end{array}
\;.
$$

The rest of the argument is like that for Theorem 4.1.

Theorem 4.3. *If N is obtained from M by interchanging columns j and j', then* $|N| = -|M|$.

Proof. Take any (M, σ, τ) with $\tau_1 = j$, $\tau_2 = j'$ and let $\tau' = \tau$ except $\tau'_1 = \tau_2$, $\tau'_2 = \tau_1$. Then clearly (N, σ, τ') exists, indeed, N_k is just M_k with columns j, j' switched. Thus $p_k(N, \sigma, \tau') = p_k(M, \sigma, \tau)$. As for powers of -1, we get (16), for we are dealing with case (1) of that proof.

Of course, analogous theorems hold for row operations. For example

Theorem 4.4. *If N is obtained from M by multiplying row i by c. Then $|N| = c|M|$.*

Proof. Assuming M invertible, row i cannot be all zeros, and so some (M, σ, τ) exists with $\sigma_1 = i$. Applying the first pivot of this to N gives

$$
\begin{array}{|cc|}
c\,p^* & c\,q \\
\\
r & s
\end{array}
\quad\longrightarrow\quad
\begin{array}{|cc|}
1/(c\,p) & q/p \\
\\
-r/(c\,p) & s - q\,r/p
\end{array}\;.
$$

The rest is as for Theorem 4.1.

Theorem 4.5. *If M^{-1} exists, $|M^{-1}| = |M|^{-1}$.*

Proof. Take any (M, σ, τ) and run it backwards, i.e., $M^{-1}, M_n, \ldots, M_2,$ M. Since the border vectors are exchanged pair by pair in this order too, we have an inversion sequence $(M^{-1}, \bar\sigma, \bar\tau)$. In light of (9), the pivot entries for this sequence are just the reciprocals of those for (M, σ, τ), that is, $p_k(M^{-1}, \bar\sigma, \bar\tau) = 1/p_{n+1-k}(M, \sigma, \tau)$. All that is left to show is

$$\prod(-1)^{s_k + t_k} = \prod(-1)^{\bar s_k + \bar t_k}. \tag{20}$$

The positions of the pivots in both inversion sequences are the same, namely (σ_k, τ_k), $1 \le k \le n$. Consider the permutation matrix P with ones in just these positions. Clearly, (P, σ, τ) and $(P, \bar\sigma, \bar\tau)$ both exist, and every $p_k = 1$ in both, so both sides of (20) equal $|P|$.

   Usually Theorem 4.5 is proved as a corollary to the product rule $|MN| = |M||N|$. We could do this too, since Theorems 4.1–4.3 provide all one needs to prove the product rule. However, we wished to show how the pivotal definition makes Theorem 4.5 intuitively clear: $|M^{-1}|$ should equal $|M|^{-1}$ since computing $|M^{-1}|$ amounts to reversing all the pivots.

   As for the product rule itself, this can also be proved directly by pivotal ideas. One generalizes inversion sequences to non-square matrices and refines the arguments of Theorem 3.3 to show that the signed product of pivots is still independent of the particular sequence

used. One then applies this result to two sequences starting with the
$n \times 2n$ matrix

$$X \quad \boxed{\begin{array}{cc} M & \quad MN \end{array}} \quad .$$
$$\qquad = Y \quad\quad = Z$$

The first sequence is obtained by inverting M to get

$$Y \quad \boxed{\begin{array}{cc} M^{-1} & \quad N \end{array}} \quad ,$$
$$\qquad = X \quad\quad = Z$$

and then inverting N to get

$$Z \quad \boxed{\begin{array}{cc} (MN)^{-1} & \quad N^{-1} \end{array}} \quad .$$
$$\qquad = X \quad\quad\quad = Y$$

The second sequence is obtained by inverting MN directly,

$$Z \quad \boxed{\begin{array}{cc} N^{-1} & \quad (MN)^{-1} \end{array}} \quad .$$
$$\qquad = Y \quad\quad\quad = X$$

As before, all is pretty easy to justify except for the matter of sign. Here
sign is particularly cumbersome—for generalized inversion sequences
σ, τ have reasonable anologs but s, t it seems do not—so let us not
describe this proof further.

5. Generalizations

Our frequent use of arrows should suggest that much of our work
can be stated in terms of graphs. Indeed, given an invertible matrix M,
let its *pivot graph* G be defined to have one vertex for each matrix (up
to order) in the combivalence class of M, and an edge between vertices

if they differ by the exchange of one pair of border vectors. Then each inversion sequence becomes a shortest path between the two "antipodal" vertices M and M^{-1} (longer paths would correspond to pivot sequences in which some border vectors are pivoted more than once). For two such paths to differ by an elementary deformation just means that they differ by one vertex. If we think of an elementary deformation as a homotopy, then Theorem 3.6 says *all shortest antipodal paths in a pivot graph are homotopic.*

Actually, there is no reason to restrict pivot graphs to invertible or square matrices. Nor is it necessary to consider only shortest or anti-podal paths. To take care of longer paths, we merely have to add two more very natural types of elementary homotopies, namely, $v_1 \ldots v_k v' v_k v_{k+1} \ldots v_n$ is homotopic to $v_1 \ldots v_k v_{k+1} \ldots v_n$; and if v_k, v_{k+1} are adjacent, $v_1 \ldots v_k v' v_{k+1} \ldots v_n$ is homotopic to $v_1 \ldots v_k v_{k+1} \ldots v_n$. Then it is not hard to show that *the pivot graph of any matrix has trivial homotopy.*

Finally, Theorem 3.6 depends only on the Exchange Axiom, and there are things other than sets of vectors that satisfy this. Indeed, there is a special name for such things.

Definition 5.1. A *matroid*, or *combinatorial pregeometry*, is a finite set E and a collection B of subsets of E, called *bases*, such that for all $B,B' \in \mathscr{B}$ and $e \in B - B'$ there exists $e' \in B' - B$ so that $B - e + e' \in \mathscr{B}$.

For an introduction to matroids, with a good bibliography, see [6]. Suffice it to say here that the notion of pivot graph of a matrix generalizes to that of a *basis graph* of a matroid, and *all basis graphs are homotopically trivial.* A precise statement and proof of this result, as well as related ideas culminating in a characterization of basis graphs, have been published by the author in [1].

Acknowledgment

I was a Ph.D. student of Professor Tucker. My thesis [2] treats various aspects of matroid basis graphs, including the generalizations just mentioned of the topics in the present paper. The pivotal theory of determinants itself does appear in the thesis, but only as a brief appendix. Yet this matter of determinants was in fact the problem Professor Tucker posed to me when he first took me on as a student. Only after

I solved it, and only after he then posed further questions, did I begin to see my way towards what became the body of the thesis.

I mention all this because it points out two significant things about Al Tucker's pedagogical and mathematical approach. First, he starts his students with something he is sure can be done. Thus they can make some early progress, which does wonders for their confidence and enthusiasm at a very crucial stage. Second, by showing himself to be interested in such "trivial" problems as defining determinants, he displays one of his strongest convictions, one that we might all take to heart, namely that significant new mathematics can arise from elementary topics.

References

[1] S.B. Maurer, "Matroid basis graphs I", *Journal of Combinatorial Theory* B 14 (1973) 216–240.
[2] S.B. Maurer, "Matroid basis graphs", Ph.D. Dissertation, Princeton University, Princeton, N.J. (May 1972).
[3] B. Noble, *Applied linear algebra* (Prentice-Hall, Englewood Cliffs, N.J., 1969).
[4] A.W. Tucker, "A combinatorial equivalence of matrices", in: *Combinatorial analysis*, Proceedings of Symposia in Applied Mathematics X, Eds. R. Bellman and M. Hall (Am. Math. Soc., Providence, R.I., 1960) pp. 129–140.
[5] A.W. Tucker, "Combinatorial theory underlying linear programs", in: *Recent advances in mathematical programming*, Eds. R.L. Graves and P. Wolfe (McGraw-Hill, New York, 1963) pp. 1–16.
[6] R.J. Wilson, "An introduction to matroid theory", *American Mathematical Monthly* 80 (1973) 505–525.

Mathematical Programming Study 1 (1974) 175–189. North-Holland Publishing Company

A NOTE ON THE LEMKE–HOWSON ALGORITHM

Lloyd S. SHAPLEY

The Rand Corporation, Santa Monica, Calif., U.S.A.

Received 15 May 1974
Revised manuscript received 8 July 1974

Dedicated, with appreciation and admiration, to A.W. Tucker

The Lemke–Howson algorithm for bimatrix games provides both an elementary proof of the existence of equilibrium points and an efficient computational method for finding at least one equilibrium point. The first half of this paper presents a geometrical view of the algorithm that makes its operation especially easy to visualize. Several illustrations are given, including Wilson's example of "inaccessible" equilibrium points. The second half presents an orientation theory for the equilibrium points of (nondegenerate) bimatrix games and the Lemke–Howson paths that interconnect them; in particular, it is shown that there is always one more "negative" than "positive" equilibrium point.

1. Introduction

The first half of this paper is frankly expository; it contains little of substance that is not covered in Lemke and Howson's original work [5]. But rather than skim over this material as rapidly as possible, we have taken this opportunity to set down the details of a geometric labeling system for bimatrix games, which we have found very effective for explaining, at least to "lay" audiences, the workings of the Lemke–Howson method.[1]

The second half presents an index theory for bimatrix games in which we show that each equilibrium point (in the nondegenerate case) has an intrinsic orientation, and each almost-complementary path or loop an intrinsic sense of direction. The possibility of such a theory will

[1] As presented, e.g., at the Second World Congress of the Econometric Society in Cambridge, England, September 1970.

The author would like to acknowledge helpful conversations and correspondence with B.C. Eaves, H.W. Kuhn, C.E. Lemke and H.E. Scarf and the support of Air Force Project RAND.

hardly surprise those who are familiar with the "strong" form of Sperner's Lemma [1, p. 133]; it is analogous to (and perhaps reducible to) Ky Fan's theory for abstract orientable pseudomanifolds given in his 1967 paper [2]. In any case, our concern here is with the concrete realization of the theory, and we shall give explicit definitions for the various indices in terms of the signs of suitably constructed determinants derived from the payoff matrices.

2. Almost-completely-labeled paths

A *bimatrix game* (A, B) consists of two *m*-by-*n* matrices

$$A = (a_{ij}: i \in I, j \in J), \qquad B = (b_{ij}: i \in I, j \in J),$$

representing the payoffs to two players using pure strategies i and j, respectively. It will be convenient to have the sets I and J disjoint, so we define

$$I = \{1, \ldots, m\}, \qquad J = \{m + 1, \ldots, m + n\}, \qquad K = I \cup J.$$

Mixed strategies are represented by vectors

$$s = (s_1, \ldots, s_m) \in S, \qquad t = (t_{m+1}, \ldots, t_{m+n}) \in T,$$

where

$$S = \{s \geqslant 0: \textstyle\sum_I s_i = 1\}, \qquad T = \{t \geqslant 0: \textstyle\sum_J t_j = 1\}.$$

The corresponding payoffs are $\sum_I \sum_J a_{ij} s_i t_j$ and $\sum_I \sum_J b_{ij} s_i t_j$. Geometrically, the sets S and T are simplices of dimension $m - 1$ and $n - 1$, respectively. Define also

$$\tilde{S} = S \cup \{s \geqslant 0: \textstyle\sum_I s_i \leqslant 1 \text{ and } \textstyle\prod_I s_i = 0\},$$
$$\tilde{T} = T \cup \{t \geqslant 0: \textstyle\sum_J t_j \leqslant 1 \text{ and } \textstyle\prod_J t_j = 0\}.$$

These extended domains are the boundaries of the simplices of the next higher dimension, "cut off" from the positive orthant by S and T, respectively.

An *equilibrium point* of (A, B) is a pair $(s^*, t^*) \in S \times T$ such that

$$\sum_I \sum_J a_{ij} s_i^* t_j^* = \max_{s \in S} \sum_I \sum_J a_{ij} s_i t_j^*,$$

$$\sum_I \sum_J b_{ij} s_i^* t_j^* = \max_{t \in T} \sum_I \sum_J b_{ij} s_i^* t_j.$$

An equivalent condition is

$$\sum_J a_{i*j} t_j^* = \max_{i \in I} \sum_J a_{ij} t_j^*, \quad \text{all } i^* \in I \text{ with } s_{i*}^* > 0,$$

$$\sum_I b_{ij*} s_i^* = \max_{j \in J} \sum_I b_{ij} s_i^*, \quad \text{all } j^* \in J \text{ with } t_{j*}^* > 0.$$

We now define certain closed convex, polyhedral regions S^k in $\tilde{S}$, as follows:

$$S^i = \{s \in \tilde{S} : s_i = 0\} \qquad \text{for } i \in I,$$
$$S^j = \{s \in S : \sum_I b_{ij} s_i = \max_{l \in J} \sum_I b_{il} s_i\} \quad \text{for } j \in J.$$

The sets S^i, $i \in I$, cover all of $\tilde{S} \backslash S$, as well as the relative boundary of S. The sets S^j, $j \in J$ (some of which may be empty) consist of those mixed strategies for the first player against which the pure strategy "j" is a best response by the second player. Since there always is at least one best response, they cover all of S. Hence the sets S^k, $k \in K$ cover all of $\tilde{S}$. Define the *label* of $s \in \tilde{S}$ to be the nonempty set

$$L'(s) = \{k : s \in S^k\}.$$

In exactly similar fashion, define regions T^k in $\tilde{T}$ by

$$T^i = \{t \in T : \sum_J a_{ij} t_j = \max_{l \in I} \sum_J a^{lj} t_j\} \quad \text{for } i \in I,$$

$$T^j = \{t \in \tilde{T} : t_j = 0\} \qquad \text{for } j \in J,$$

and, for $t \in \tilde{T}$, define the label

$$L''(t) = \{k : t \in T^k\}.$$

Finally, let the label of the pair $(s, t) \in \tilde{S} \times \tilde{T}$ be

$$L(s, t) = L'(s) \cup L''(t).$$

We shall say that (s, t) is *completely labeled* if $L(s, t) = K$, and *almost completely labeled* if $L(s, t) = K\backslash\{k\}$ for some $k \in K$.

Theorem 2.1. *If $(s, t) \in S \times T$, then (s, t) is an equilibrium point of (A, B) if and only if (s, t) is completely labeled.*

This follows almost immediately from the definitions. We now make a "nondegeneracy" assumption.

Assumption 2.2. The game (A, B) is such that
(1) each non-empty region S^k in $\tilde{S}$ is $(m - 1)$-dimensional;
(2) the intersection of any two of the S^k is at most $(m - 2)$-dimensional;
(3) no point of $\tilde{S}$ belongs to more than m of the S^k;
(4) the analogous conditions to (1), (2), (3) hold for the regions T^k in $\tilde{T}$.

As the failure of any of these conditions would entail a special numerical relationship among the a_{ij} or b_{ij}, we see that "almost all" bimatrix games are nondegenerate, in this sense.

Armed with this assumption, we can describe a *graph* $\mathscr{S}$ in $\tilde{S}$, whose point set consists of all points in $\tilde{S}$ that belong to at least $m - 1$ of the regions S^k. The points that belong to exactly m regions are the *nodes* of $\mathscr{S}$, while those that belong to exactly $m - 1$ regions make up the *edges*. In a given edge all points have the same label, which we shall consider the label of that edge. The nodes and edges of $\mathscr{S}$ exhibit the following incidence relations:
(1) each edge touches exactly two nodes (i.e., its end-points),
(2) each node touches exactly m edges, each one omitting from its label a different member of the label of the node.

We call two nodes *adjacent* if they are at opposite ends of the same edge, or, equivalently (since the regions are convex), if their labels differ in exactly one element.

An exactly analogous graph $\mathscr{T}$ can be described in $\tilde{T}$. The construction of these graphs is illustrated in Fig. 1 for $\mathscr{S}$, with $m = n = 3$, where we have taken B to be the identity matrix I_3. In the equivalent planar diagram at the right, the exceptional node 0 is at the top, and region ① is unbounded. It will be seen that the effect of our rather unusual extension of the mixed-strategy simplex from S to $\tilde{S}$ has been to "close

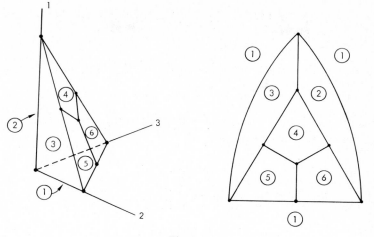

Fig. 1.

out" the graph, making it regular of degree m by adding one node and m edges.

We now turn our attention to node pairs.[2] Let $\mathcal{N}$ denote the set of all $(s, t) \in \tilde{S} \times \tilde{T}$, where s is a node of $\mathcal{S}$ and t a node of $\mathcal{T}$. Define

$$\mathcal{P} = \{(s, t) \in \mathcal{N} : L(s, t) = K\}$$

and, for each $k \in K$,

$$\mathcal{P}^k = \{(s, t) \in \mathcal{N} : L(s, t) \supseteq K \backslash \{k\}\}.$$

[2] Implicitly, we are forming a kind of "product" graph out of $\mathcal{S}$ and $\mathcal{T}$, in which the node-node pairs from $\mathcal{S}$, $\mathcal{T}$ are the nodes and the node-edge and edge-node pairs are the edges.

The "graph" approach to path-following algorithms is more or less *dual* to the "pseudo-manifold" (or regular simplicial complex) approach. Thus, the nodes and edges of $\mathcal{S}$ and the $(m - 1)$-dimensional regions S^k correspond to the simplices of dimension $m - 1$, $m - 2$ and 0, respectively, of the simplicial complex that is dual to the polyhedral complex in $\tilde{S}$ of which $\mathcal{S}$ is the regular, 1-dimensional skeleton. Kuhn [3] suggests viewing the Lemke–Howson method as operating on a product of disjoint pseudomanifolds, which is itself trivially a pseudomanifold. However, the success of the method does not depend on $\mathcal{S}$ actually being the skeleton of the dual of a pseudomanifold, since no conditions need be satisfied by the cells of intermediate dimension between 1 and $m - 1$, nor need the regions S^k be convex or even connected. In an unpublished 1970 note, the present author proposed the use of combinatorially defined objects called "regular quasi-skeletons", as a generalized setting for the Lemke–Howson method. But it is not clear that RQS's not arising from orientable pseudomanifolds would support any kind of orientation theory.

These are the node pairs that are completely or almost completely labeled; note that $\mathscr{P}^k \supseteq \mathscr{P}$, and that $k \neq l$ implies $\mathscr{P}^k \cap \mathscr{P}^l = \mathscr{P}$. From Theorem 2.1 and our nondegeneracy assumption, it is easy to see that the members of $\mathscr{P}$ are just the equilibrium points of (A, B) together with the node pair $(0, 0)$.

Let us call two members of $\mathscr{N}$ *adjacent* if their $\mathscr{S}$-components are the same and their $\mathscr{T}$-components adjacent, or their $\mathscr{T}$-components are the same and their $\mathscr{S}$-components adjacent. A subset $\mathscr{R}$ of $\mathscr{N}$ is said to be *connected* if a chain of members of $\mathscr{R}$ can be found, each one adjacent to the next, joining any two given members of $\mathscr{R}$. A *component* of a subset $\mathscr{R}$ of $\mathscr{N}$ is a maximal connected subset of $\mathscr{R}$; every subset of $\mathscr{N}$ is the disjoint union of its components. A nonempty connected set $\mathscr{R}$ is called a *loop* if every member of $\mathscr{R}$ is adjacent to exactly two other members of $\mathscr{R}$, and a *path* if every member is adjacent to exactly two others except for two members (i.e., the endpoints) that are adjacent to just one each. A loop has at least three members, a path at least two.[3]

Lemma 2.3. *Let $k \in K$ be fixed.*
   (a) *Each member of $\mathscr{P}$ is adjacent to exactly one member of $\mathscr{P}^k$.*
   (b) *Each member of $\mathscr{P}^k \backslash \mathscr{P}$ is adjacent to exactly two members of $\mathscr{P}^k$.*

Proof. (a) Take a completely labeled pair $(s, t) \in \mathscr{P}$ and select the node $r, r = s$ or t, whose label includes k. Following the edge out of r whose label omits k yields a member of $\mathscr{P}^k$ adjacent to (s, t); this is uniquely determined, since any other edge out of s or t would take us out of $\mathscr{P}^k$.

   (b) Take (s, t) in $\mathscr{P}^k \backslash \mathscr{P}$ and note that there must be exactly one duplication in the label, say $\{h\} = L'(s) \cap L''(t)$. Following the edge out of s that omits h yields one adjacent member of $\mathscr{P}^k$, and following the edge out of t that omits h yields another; they are clearly distinct, and there are no others.

From Lemma 2.3, it follows that each component of $\mathscr{P}^k$ is either a path or a loop, and that the path endpoints are precisely the members of $\mathscr{P}$. Hence $\mathscr{P}$ has an even number of members, and since $(0, 0)$ is not an equilibrium point, we have:

---

[3] In our application, however, we shall find that paths of length two and loops of length three (or any odd length) cannot occur.

Theorem 2.4 (Lemke–Howson). *Every nondegenerate bimatrix game has an odd number of equilibrium points (and hence at least one).*

3. Examples

To exploit the "constructive" aspect of this existence proof, we need to be able to place ourselves on a path component of some $\mathscr{P}^k$. This is actually very easy, since the exceptional node pair $(0,0)$ belongs to $\mathscr{P}$, and hence to every $\mathscr{P}^k$. Moreover, it is never on a loop. Thus, assuming we know how to read labels and follow edges in $\mathscr{S}$ and $\mathscr{T}$, all we have to do to find an equilibrium point is select a value for k, start at $(0,0)$, and follow the path to its end.

In working through the examples, the reader may enjoy placing two small checkers or coins on the page, one on $\mathscr{S}$ and one on $\mathscr{T}$, and then making alternate moves. The "state of the system" at any given time is given by the positions of the two coins, plus an indication of which one moved last.

Let us try to explore $\mathscr{P}^1$ in Fig. 2. Starting at the exceptional node-pair (coins on "O" and "o"), we find that we must first make a move in $\mathscr{S}$,

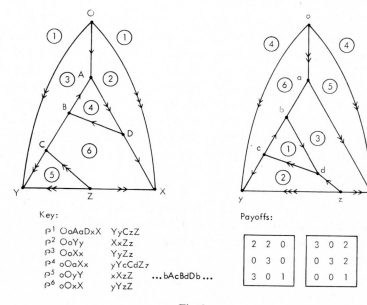

Key:

$\mathscr{P}^1$ OoAaDxX YyCzZ
$\mathscr{P}^2$ OoYy XxZz
$\mathscr{P}^3$ OoXx YyZz
$\mathscr{P}^4$ oOaXx yYcCdZz
$\mathscr{P}^5$ oOyY xXzZ ...bAcBdDb...
$\mathscr{P}^6$ oOxX yYzZ

Payoffs:

2	2	0
0	3	0
3	0	1

3	0	2
0	3	2
0	0	1

Fig. 2

sliding the coin along the edge that leads away from region ① (the unbounded, "outside" region as we have drawn it). We thereby arrive at "*A*", only to find that ④ is now doubly represented, in the label of the node pair (*A*, *o*). Moving away from ④ in $\mathscr{T}$ takes us to "*a*", where we pick up contact with ③. We therefore move in $\mathscr{S}$ from "*A*" to "*D*", picking up ⑥; then in $\mathscr{T}$ from "*a*" to "*x*", picking up ④ again; and finally, in $\mathscr{S}$, from "*D*" to "*X*", picking up ① for a complete label. So the pair (*X*, *x*) is an equilibrium point, in which each player uses just his third pure strategy.

The $\mathscr{P}^2$ path leading out of (*O*, *o*) is shorter, containing just the two other pairs (*Y*, *o*) and (*Y*, *y*). The latter is another equilibrium point of the game, with each player using just his second pure strategy. This gives us the opportunity to start another $\mathscr{P}^1$ path, which will necessarily lead us to a third equilibrium point. In fact, moving away from ① at "*Y*" takes us to "*C*", duplicating ⑥ in the label; moving away from ⑥ at "*y*" takes us to "*z*", duplicating ③; and finally, moving away from ③ at "*C*" takes us to the equilibrium point (*Z*, *z*). This equilibrium point involves a mixture of the second and third strategies for each player. As it happens, there are no other equilibrium points.[4]

The "Key" in Fig. 2 lists all of the $\mathscr{P}^k$'s, each "word" representing a different component, with the completely-labeled or almost completely-labeled node pairs indicated by adjacent letters. Especially interesting is the third component of $\mathscr{P}^5$, which is a loop of length 6. Note that there are no paths from (*O*, *o*) to (*Z*, *z*) or from (*X*, *x*) to (*Y*, *y*); the reason, as we shall see in the next section, is that both the former have index $+1$ and both the latter -1, while according to Theorem 4.5 the endpoints of a path always have indices of opposite sign.[5]

The loop in $\mathscr{P}^5$ in Fig. 2 could not have been discovered by simple path-following. (Of course, in an example of this size, an exhaustive search is easy.) Fig. 3 illustrates that inaccessible *paths* can also occur. Indeed, an inspection of the "Key" reveals six ways to get from (*O*, *o*) to (*X*, *x*) and six ways to get from (*Y*, *y*) to (*Z*, *z*), but no interconnection.[6]

---

[4] The maximum number of equilibrium points for a nondegenerate 3×3 bimatrix game is seven; this may be attained by taking both *A* and *B* to be the identity matrix. We do not know if the maximum is achieved similary in larger games; the question is related to the question of determining theoretical upper bounds for Lemke–Howson path lengths.

[5] For an explanation of the arrows in Figs. 2 and 3, see the remarks preceding Theorem 4.6.

[6] The example is due to Robert Wilson (private correspondence, 1970).

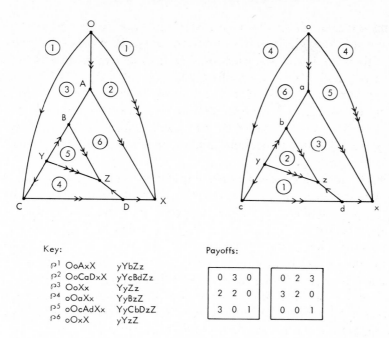

Key:

$\wp^1$ OoAxX   yYbZz
$\wp^2$ OoCaDxX   yYcBdZz
$\wp^3$ OoXx   YyZz
$\wp^4$ oOaXx   YyBzZ
$\wp^5$ oOcAdXx   YyCbDzZ
$\wp^6$ oOxX   yYzZ

Payoffs:

$$\begin{array}{ccc} 0 & 3 & 0 \\ 2 & 2 & 0 \\ 3 & 0 & 1 \end{array} \qquad \begin{array}{ccc} 0 & 2 & 3 \\ 3 & 2 & 0 \\ 0 & 0 & 1 \end{array}$$

Fig. 3

This shows that, in general, even the most thoroughgoing exploration of the path network radiating from the starting point (O, o) may not find all solutions, or even tell us whether there are any solutions unfound. This is a serious practical drawback to the Lemke–Howson method, which is in other respects quite efficient computationally. The multiplicity of solutions is admittedly a weakness of the equilibrium-point solution concept, but there does not appear to be any good reason heuristically to prefer the solutions that are accessible to path-following, or to reject those that are not. An efficient method for finding *all* equilibrium points in bimatrix games would be an important contribution.

It is rather surprising that the 3×3 framework should have room for so much interesting behavior. But the apparent simplicity of the setting is deceptive; there are 64 node pairs in $\mathcal{N}$ in any nondegenerate 3×3 game if none of the regions S^k, T^k are empty. In our two examples, about half of the 64 are actually used in tracing the almost-completely-labeled paths and loops; in larger games the proportion used would presumable be much smaller.

4. Index theory

Given $(s, t) \in \tilde{S} \times \tilde{T}$, define the *index matrix* of (s, t) to be the $(m + n)$-by-$(m + n)$ matrix $C = (c_{kl})$ with entries

$$c_{kl} = \begin{cases} 1 & \text{if } k = l \in I \cap L'(s), \\ b_{kl} & \text{if } k \in I \text{ and } l \in J \cap L'(s), \\ a_{lk} & \text{if } k \in J \text{ and } l \in I \cap L''(t), \\ 1 & \text{if } k = l \in J \cap L''(t), \\ 0 & \text{otherwise.} \end{cases}$$

This is illustrated for $m = 3$, $n = 2$, $L'(s) = \{1, 3, 5\}$ (dots), and $L''(t) = \{3, 4\}$ (crosses):

	•		•		•
I	1	0	0	0	b_{15}
	0	0	0	0	b_{25}
	0	0	1	0	b_{35}
J	0	0	a_{34}	1	0
	0	0	a_{35}	0	0
			×	×	

In this case, with only the second column all zero, we have $(s, t) \in \mathscr{P}^2$.

Assume for the moment that all a_{ij}, b_{ij} are positive.[7] The *index* of (s, t) is then defined to be the sign of the determinant of the index matrix of (s, t), that is, a number 1, -1, or 0. We note first that the index does not depend on the order in which the players or their pure strategies are numbered. It is obvious that any pair (s, t) *not* in $\mathscr{P}$ will have index 0. Moreover, under our nondegeneracy assumption, no element of $\mathscr{P}$ will have index 1 or -1. Our main object will be to prove that the indices of the endpoints of any path in any $\mathscr{P}^k$ are of opposite sign.

First we shall prove a lemma that "explains" the positivity assumption in the definition of the index.

---

[7] See the remark immediately following the proof of Lemma 4.1.

Lemma 4.1. *Let (A, B) be a nondegenerate bimatrix game with $A > 0$, $B > 0$. Form a new game by adding nonnegative constants x and y to all the elements of A and B, respectively. Then L', L'', $\mathcal{P}$, $\mathcal{S}$, $\mathcal{T}$, and the $\mathcal{P}^k$, $k \in K$ are all unchanged, as are the indices of all $(s, t) \in \tilde{S} \times \tilde{T}$.*

Proof. That the labels, graphs, and zero indices are unchanged is immediate from the definitions. With nondegeneracy, the determinant of the index matrix for $(s, t) \in \mathcal{P}$ reduces, because of the blocks of zeros, to

$$\pm (\det \bar{A})(\det \bar{B}),$$

where $\bar{A}$ and $\bar{B}$ are the square submatrices associated with the positive parts of s and t and where the choice in the "$\pm$" symbol depends only on the labels of s and t. It will suffice to show that adding a positive constant does not change the sign of (say) $\det \bar{B}$. But $\bar{B}$ has the property that left-multiplication by $\bar{s}$ yields a constant, i.e.,

$$\bar{s}\bar{B} = (v, v, \ldots, v)$$

for some $v > 0$, where we have written $\bar{s}$ for the subvector made up of the positive components of s. Define $f(x)$ to be the determinant of the matrix $\bar{B}$ with each entry increased by x. Then $f(-v) = 0$. Since f is linear in x,[8] it follows that the sign of $f(x)$ is constant for $x \geqslant 0$. This completes the proof of Lemma 4.1.

In view of this lemma, it makes sense to define the indices for an arbitrary nondegenerate bimatrix game by adding constants to convert it to an equivalent positive game.

Now consider the m-by-$(m + n)$ matrix

$$[-I_m, B] = \begin{array}{|ccccc|cccc|} \hline -1 & 0 & \ldots & 0 & & b_{1,m+1} & \ldots & b_{1,m+n} \\ 0 & -1 & \ldots & 0 & & b_{2,m+1} & & \vdots \\ & \ldots & & & & \ldots & & \\ 0 & & \ldots & -1 & & b_{m,m+1} & \ldots & b_{m,m+n} \\ \hline \end{array}$$

---

[8] To see this, subtract one column of the matrix from all the other columns.

Given an arbitrary node s of $\mathscr{S}$, we can use the label $L'(s)$ to select columns of $[-I_m, B]$ to form a square matrix $B(s)$. Except for the minus signs, $B(s)$ will consist of the nonzero columns of the upper part of the index matrix of (s, t), for any t.

Lemma 4.2 *Let s' be adjacent to s in $\mathscr{S}$, and let the columns of $B(s)$ and $B(s')$ be so ordered that the two matrices are identical except in one column. Let $B > 0$. Then the determinants of $B(s)$ and $B(s')$ are opposite in sign.*

This sort of lemma is familiar in linear programming,[9] but we do not have here quite the standard set-up, so we shall prove it afresh.

Proof of Lemma 4.2. Let s be a node in $\mathscr{S}$, and let $v = \max_J \sum_I b_{ij} s_i$; the latter is positive because $B > 0$. Define the vector p by the multiplication

$$(s_1, \ldots, s_m)[-I_m, B] = (p_1, \ldots, p_{m+n}),$$

and define $\delta_{Jk} = 0$ for $k \in I$, $\delta_{Jk} = 1$ for $k \in J$. Then we have

$$p_k \begin{Bmatrix} = \\ < \end{Bmatrix} \delta_{Jk} v \quad \text{for } k \begin{Bmatrix} \in \\ \notin \end{Bmatrix} L'(s).$$

Similarly, defining v' and p' analogously for the adjacent node s', we have

$$p'_k \begin{Bmatrix} = \\ < \end{Bmatrix} \delta_{Jk} v' \quad \text{for } k \begin{Bmatrix} \in \\ \notin \end{Bmatrix} L'(s').$$

Let k^* and l^* be the columns of $[-I_m, B]$ that make the difference between $B(s)$ and $B(s')$. That is, let

$$\{k^*\} = L'(s)\backslash L'(s'), \quad \{l^*\} = L'(s')\backslash L'(s).$$

Then there are positive numbers α, α' such that

$$p_{l^*} = \delta_{Jl^*} v - \alpha \quad \text{and} \quad p'_{k^*} = \delta_{Jk^*} v' - \alpha'.$$

[9] When one pivots only on negative entries, the determinant of the "current basis matrix," which is the product of pivots to date, changes sign at each step.

For $0 \leqslant \lambda \leqslant 1$, define $B_\lambda = (1 - \lambda) B(s) + \lambda B(s')$. Multiplying $v's - v s'$ into B_λ, we obtain all zeros except for the inner product of $v's - v s'$ with the replacement column; there we obtain

$$v'[(1 - \lambda) p_{k*} + \lambda p_{l*}] - v[(1 - \lambda) p'_{k*} + \lambda p'_{l*}] =$$
$$= v'[(1 - \lambda)\delta_{Jk*}v + \lambda(\delta_{Jl*}v - \alpha)]$$
$$- v[(1 - \lambda)(\delta_{Jk*}v' - \alpha') + \lambda \delta_{Jl*}v']$$
$$= -v'\lambda \alpha + v(1 - \lambda) \alpha'$$
$$= v \alpha' - \lambda(v \alpha' + v'\alpha).$$

Thus, $v's - v s'$ annihilates the matrix B_λ when $\lambda = v \alpha'/(v \alpha' + v'\alpha)$. This number lies between 0 and 1 since v, v', α, α' all are positive. So the determinant of B_λ, which is linear in λ and not identically 0, has opposite signs at $\lambda = 0$ and $\lambda = 1$. This completes the proof of Lemma 4.2.

Now let (s, t) be an almost completely labeled node pair, say $(s, t) \in \mathscr{P}^k \backslash \mathscr{P}$, and let h be the duplication in its label: $\{h\} = L'(s) \cap L''(t)$. Assuming $A > 0, B > 0$, define the $\mathscr{S}$-index of (s, t) to be the sign of the determinant of the "$\mathscr{S}$-index matrix" of (s, t), obtained from the index matrix of (s, t) by a partial column transposition, namely, interchange the "upper" (or "I") portion of the kth column (which contains all zeros) with the upper portion of the hth column (which is the hth unit vector if $h \in I$ or the hth column of B if $h \in J$). The $\mathscr{T}$-index of (s, t) is defined analogously by making a transposition in the lower or "J" portion of the index matrix. By non-degeneracy and positivity, both of these indices are nonzero.

Lemma 4.3. *The $\mathscr{S}$-index and the $\mathscr{T}$-index for a given $(s, t) \in \mathscr{P}^k \backslash \mathscr{P}$ have opposite signs.*

Proof. The defining matrices differ only by a single column transposition.

For completely labeled node pairs it will be convenient to define both the $\mathscr{S}$-index and the $\mathscr{T}$-index to be equal to the *index*, as previously defined.

Lemma 4.4. *Let (s, t) and (s', t) be adjacent node pairs in $\mathscr{P}^k$. Then their $\mathscr{S}$-indices have opposite signs.*

Proof. Regardless of whether one of (s, t), (s', t) is in $\mathscr{P}$ or not, their $\mathscr{S}$-index-defining matrices differ only in the "upper" portions of their k^{th} columns, which reflect the replacement step between s and s'. Thus, the determinants of these matrices have the form $\pm(\det B(s))(\det \overline{A})$ and $\pm(\det B(s'))(\det \overline{A})$, with the *same* choice in the "$\pm$" symbol and with $B(s)$ and $B(s')$ related as in Lemma 4.2. Hence their signs are opposite.

We can now define a "sense of direction" on the paths and loops of $\mathscr{P}^k$. Arbitrarily, let a "forward" step at (s, t) be a move to an adjacent member of $\mathscr{P}^k$ that changes s only if the $\mathscr{S}$-index of (s, t) is positive, or t only if the $\mathscr{T}$-index of (s, t) is positive. Similarly, a "backward" step changes s only if the $\mathscr{S}$-index is negative and t only if the $\mathscr{T}$-index is negative. Hence every move to an adjacent member of $\mathscr{P}^k$ is either a forward or a backward step, and by Lemma 4.3 we have, at each *almost*-completely-labeled node pair, the choice between a forward and a backward step. Moreover, by Lemma 4.4, the reversal of any forward step is a backward step, and vice versa. Hence we have

Theorem 4.5. *Moving forward along any path in any $\mathscr{P}^k$ leads to an equilibrium point with index -1. Moving backward along any path in any $\mathscr{P}^k$ leads to an equilibrium point (or the special point $(0, 0)$) with index $+1$. Hence there is exactly one more equilibrium point with index -1 than with index $+1$.*

What we have done in the above has been to establish an orientation[10] on certain edges in the "product" graph of $\mathscr{S}$ and $\mathscr{T}$, namely those edges that carry the paths and loops of the $\mathscr{P}^k$, $k \in K$. We might also attempt to orient the edges of the separate graph $\mathscr{S}$ (and similarly for $\mathscr{T}$), by ascribing a "forward" sense to the directed edge $<s, s'>$ whenever the directed edge $<(s, t), (s', t)>$ represents a forward step in $\mathscr{P}^k$, for some t and k. Rather surprisingly, this works. The same edge $<s, s'>$ may well participate in several different paths or loops, but, as Theorem 4.6 will show, all the orientations induced on it will agree. These induced orientations are illustrated by the multiple arrows in Figs. 2 and 3.[11]

[10] Compare Kuhn's observations [4] on oriented Sperner graphs.

[11] As it happens, almost every edge shown receives at least one arrowhead; this is because m and n are so small. If we interpret the arrowheads as flow units, then the solution nodes appear as sources and sinks (index $+1$ and -1, respectively), while the other nodes balance inflow and outflow.

Theorem 4.6. *Let s and s′ be adjacent nodes in $\mathscr{S}$, and let t, k, t′, k′ be such that (s, t) and (s′, t) are in $\mathscr{P}^k$ and (s, t′) and (s′, t′) are in $\mathscr{P}^{k'}$. Then the $\mathscr{S}$-indices of (s, t) and (s, t′) are equal. Hence the move from (s, t) to (s′, t) in $\mathscr{P}^k$ and the move from (s, t′) to (s′, t′) in $\mathscr{P}^{k'}$ are either both forward steps or both backward steps.*

Proof. If $t = t'$, the result is trivial. If $t \neq t'$ (and hence $k \neq k'$) we first show that t and t' are adjacent. Indeed, by definition of $\mathscr{P}^k$ we have

$$L''(t) \cup [L'(s) \cap L'(s')] \cup \{k\} = K,$$

and since the three terms in this union have cardinality n, $m - 1$ and 1, respectively, the union is disjoint. Similarly, the union

$$L''(t') \cup [L'(s) \cap L'(s')] \cup \{k'\} = K$$

is disjoint, and so we see that the labels of t and t' differ in only one element:

$$L''(t) \backslash L''(t') = \{k'\}; \qquad L''(t') \backslash L''(t) = \{k\}.$$

Hence t and t' are adjacent as claimed. It follows that the matrices $A(t)$ and $A(t')$, formed like $B(s)$ and $B(s')$ in Lemma 4.2, have oppositely-signed determinants.

If we now consider the defining matrices for the $\mathscr{S}$-indices of (s, t) and (s, t') (whether or not one of them is in P), we see that their determinants have the form $\pm(\det B(s))(\det A(t))$ and $\pm(\det B(s))(\det A(t'))$, with *opposite* choices in the two "$\pm$" symbols this time, because columns k and k' must be transposed. Since $\det A(t)$ and $\det A(t')$ also have opposite signs, the two indices are equal.

References

[1] S. S. Cairns, *Introductory topology* (Ronald Press, New York, 1961).
[2] Ky Fan, "Simplicial maps from an orientable *n*-pseudo-manifold into S^m with the octahedral triangulation", *Journal of Combinatorial Theory* 2 (1967) 588-602.
[3] H. W. Kuhn, "Systematic search on pseudomanifolds", *SIAM Journal on Control*, to appear.
[4] H. W. Kuhn, "A new proof of the fundamental theorem of algebra", *Mathematical Programming Study* 1 (1974) 148-158.
[5] C. E. Lemke and J. T. Howson, Jr., "Equilibrium points of bimatrix games", *Journal of the Society for Industrial and Applied Mathematics* 12 (1964) 413-423.

Mathematical Programming Study 1 (1974) 190–205. North-Holland Publishing Company

ALGORITHM FOR A LEAST-DISTANCE
PROGRAMMING PROBLEM

Philip WOLFE

IBM Thomas J. Watson Research Center, Yorktown Heights, N.Y., U.S.A.

Received 4 March 1974
Revised manuscript received 25 March 1974

An algorithm is developed for the problem of finding the point of smallest Euclidean norm in the convex hull of a given finite point set in a Euclidean space, with particular attention paid to the description of the procedure in geometric terms.

Dedication

The paper that follows was written in celebration of the work of Albert W. Tucker in mathematical programming. It also bears witness to his pervasive influence on my own work; the three postdoctoral years I spent working under his direction at Princeton set the aims and style of much of the rest of my career, and I have called on his helpfulness and wisdom many times since then.

The paper deals with a certain quadratic programming problem and its solution via pivoting in a condensed tableau. Al Tucker pioneered in these areas, and though his first work in them was done more than twenty years ago and mine just a little later, they have continued to involve us both.

To give a highly condensed account of the development of "quadratic programming", Al's papers [14, 15, 11] are the earliest I have found which point out the usefulness and attractive properties of the quadratic programming problem, and I believe it was these papers that led Edward Barankin (my thesis advisor at Berkeley) and Robert Dorfman to make the important connection between the simplex tableau and the Kuhn—Tucker conditions for that problem [1, 2]. At Princeton, in Al's ONR

Logistics Project, Marguerite Frank and I used their result as an essential part of applying what has become known as the Frank–Wolfe algorithm to the quadratic programming problem [10]; later, still with the Project, I developed it further into "A simplex method for quadratic programming" [20], which was the first use of complementary bases in a nonlinear programming algorithm.

Meanwhile, Al continued on one hand to exploit tableaux and pivot theory in many ways eg. [16, 17], and on the other hand to analyze the structure of the general quadratic programming problem through the particularly graphic properties of his "least-distance programming problem", in which the minimand is the square of the Euclidean metric [18, 19, 13]. The present paper is about a special case of that problem, and aims at relating, for an algorithm, the geometry of the problem—which Al has used in many beautiful ways—to the algebra of its solution by the tableaux Al taught me to use.

I hope he likes it.

1. Introduction

We develop a numerical algorithm for finding that point of a polytope in Euclidean *n*-space having smallest Euclidean norm. The polytope is defined as the convex hull of a given point set $P = \{P_1, \ldots P_m\}$; algebraically, we are required to minimize $|X|^2 = X^T X$ for all X of the form

$$X = \sum_{k=1}^{m} P_k w_k, \quad \sum_{k=1}^{m} w_k = 1 \quad \text{for all } w_k \geq 0. \tag{1.1}$$

This problem is, of course, a problem of quadratic programming, for which there are several excellent general algorithms [3, 5, 9, 21] which would probably all behave with comparable efficiency; but the special nature of this problem—particularly the uniqueness of its solution X—encourages a more refined attack on it. The problem arises in a surprising number of places, if one looks for it: we have used its solution in a procedure for the minimization of nondifferentiable functions and in the construction of hyperplanes separating two given point sets as in [4].

Since problem (1.1) is very much a geometry problem, we have taken the route of first devising a plausible "geometric" procedure for solving it, described in Section 2 and illustrated on a small example, and then

using the familiar algebra of complementary pivot theory to describe
the calculations. The resulting algorithm (as explained in Section 7) may
be somewhat more efficient than some of the standard algorithms when
applied to this problem, but our primary interest here is in the fact that
all of the quantities that are consulted in performing it are geometrically
explainable. Sections 3 and 4 are devoted to the connections between
the entries of the tableau used in the calculation and the geometric
elements of importance. The algebra of the algorithm is set forth in
Section 5, while Section 6 tracks it in solving the example. Section 7 deals
with some numerical safeguards needed to make the method work on
a real computer, results from running some test problems, and the
conditions under which one would want to use this algorithm in practice.

The last section also refers to a different way [22] of organizing the
calculations of the algorithm which we find better than this tableau form,
except when $n > m$. That was the original form of the algorithm; we are
indebted to Harlan Crowder for the suggestion of casting it into tableau
form, which was particularly appropriate for a problem of his in which
$n = 500$ and $m = 10$.

The following expressions are used frequently below: For a set Q of k
points in E^n (taken to be a set of k column vectors, or the columns of the
n-by-k matrix Q),

$$A(Q) = \{X : X = Q w, e^T w = 1\}$$

is the affine hull of Q, and

$$C(Q) = \{X : X = Q w, e^T w = 1, w \geq 0\}$$

is the convex hull of Q. [e is the column vector $(1, 1, \ldots, 1)^T$. The number
of components in vectors like e and w is to be inferred from the context:
here, k]. The set Q is affinely independent if $q \in A(Q \setminus \{q\})$ is false for all
$q \in Q$. Finally, for any $X \in E^n$

$$H(X) = \{Y \in E^n : X^T Y = X^T X\}$$

will denote the ($n - 1$)-dimensional affine set passing through the point
X and normal to the line through X and the origin.

2. The algorithm: geometry

We call an affinely independent subset Q of P a *corral* if the point of smallest norm in $C(Q)$ is in its relative interior. Note that any singleton is a corral. There is a corral whose convex hull contains the solution of the smallest-norm problem over P, and our algorithm will find it.

The algorithm consists of a finite number of *major cycles*, each of which consists of a finite number of *minor cycles*. At the start of each major cycle we have a corral Q and the point $X \in C(Q)$ of smallest norm. At the start of a minor cycle we have merely some affinely independent set Q and some point $X \in C(Q)$. Each minor cycle yields another such Q and X until the last, which finds a corral; that major cycle is finished, and another ready to begin. The initial corral is the singleton given by step 0 below. Subsequently, each major cycle begins at step 1; the minor cycles, if any, in a major cycle constitute repetitions of steps 3 and 2.

Algorithm

Step 0. Find a point of P of minimal norm. Let X be that point and $Q = \{X\}$.

Step 1. If $X = 0$ or $H(X)$ separates P from the origin, stop. Otherwise, choose $P_J \in P$ on the near side of $H(X)$ and replace Q by $Q \cup \{P_J\}$.

Step 2. Let Y be the point of smallest norm in $A(Q)$. If Y is in the relative interior of $C(Q)$, replace X by Y and return to step 1. Otherwise

Step 3. Let Z be the nearest point to Y on the line segment $C(Q) \cap XY$ (thus a boundary point of $C(Q)$). Delete from Q one of the points not on the face of $C(Q)$ in which Z lies, and replace X by Z. Go to step 2.

The example below illustrates the algorithm in a simple problem, giving the current Q and X at the end of each step.

We must show that the algorithm terminates in a solution of the problem. First, observe that Q is always affinely independent: it changes only by the deletion of single points or by the adjunction of P_J in step 1. Now the line $\overline{OX}$ is normal to $A(Q)$, since $|X|$ is minimal there; thus $A(Q) \subseteq H(X)$. Since $P_J \notin H(X)$, we know $P_J \notin A(Q)$, so $Q \cup \{P_J\}$ is affinely independent if Q is. Next, there can be no more minor cycles in a major cycle beginning with a given Q than the dimension of $C(Q)$, for when Q is a singleton, step 2 returns us to step 1 (indeed, the total number of minor cycles that have been performed from the beginning cannot exceed the number of major cycles). Every time step 1 is followed by the replacement in step 2 (the major cycle has no minor cycle) the value of $|X|$ is reduced, since the segment XY intersects the interior of $C(Q \cup \{P_J\})$, and $|X|$ strictly decreases along that segment. For the same reason, the first minor cycle, if any, of a major cycle also reduces $|X|$, and subsequent minor cycles cannot increase it. Thus $|X|$ is reduced in each major cycle. Since X is uniquely determined by the corral on hand at step 1, no corral can enter the algorithm more than once. Since there is but a finite number of corrals, the algorithm must terminate, and it can only do so when the problem is solved.

The reader familiar with the Simplex Method for linear programming will notice its close relationship to our algorithm, particularly in Dantzig's description [7, Section 7–3] emphasizing the role of simplices in visualizing his method.

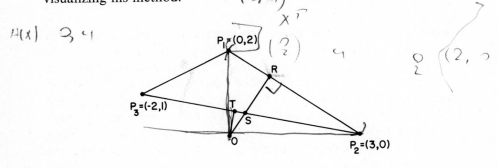

Fig. 1.

Example (see Fig. 1). $R = (0.923, 1.385)$ is the nearest point to O on P_1P_2. $S = (0.353, 0.529)$ is the intersection of OR and P_2P_3. $T = (0.115, 0.577)$ is the answer.

Here are the steps taken, and the results:

Step	X	Q	Y
0	P_1	P_1	
1	P_1	P_1, P_2	
2	R	do	R
1	R	P_1, P_2, P_3	
2	R	do	O
3	S	P_2, P_3	
2	T	do	T
1		Stop	

3. Calculating with affine sets

Affine independence of a set Q of k points is equivalent to the property that the $(n + 1)$-by-k matrix

$$\begin{bmatrix} e^T \\ Q \end{bmatrix}$$

has rank k, as well as to the property that the symmetric matrix of order $k + 1$

$$\begin{bmatrix} 0 & e^T \\ e & Q^TQ \end{bmatrix} \tag{3.1}$$

be nonsingular. (To prove the latter, suppose there were μ, u so that $e^Tu = 0$, $e\mu + Q^TQ\,u = 0$. Then $0 = u^T(e\mu + Q^TQ\,u) = |Q\,u|^2$, so $Q\,u = 0$, whence Q is affinely dependent.) The sets $A(Q)$ and $C(Q)$ then have dimension $k - 1$, and $C(Q)$ is a nondegenerate simplex whose vertices are the points of Q, while all the faces of dimension p of that simplex are the convex hulls of all the subsets of $p + 1$ points of Q. Also, the smallest face of $C(Q)$ containing a given point X in its relative interior is uniquely defined: its vertices are those points for which the barycentric coordinate w_j of $X = Q\,w$ in Q is positive.

When Q is affinely independent, we can find the projection X of any point Y on $A(Q)$, that is, solve the problem

$$\begin{aligned} &\text{minimize } |Y - X|^2 = |Y - Q\,u|^2, \\ &\text{subject to } e^Tu = 1. \end{aligned}$$

The gradient in u of the Lagrangian $\frac{1}{2}(Y - Q\,u)^T(Y - Q\,u) + \mu(e^Tu - 1)$ being $-Q^TY + Q^TQ\,u + e\,\mu$, we obtain the Lagrangian conditions

$$\begin{aligned} e^Tu &= 1 \\ e\mu + Q^TQ\,u &= Q^TY, \end{aligned} \tag{3.2}$$

which have a unique solution owing to the nonsingularity of their matrix (3.1). The set Q is a corral when those equations, for $Y = 0$, have a solution $u > 0$.

4. Geometry of the tableau

The problem of this paper is:

$$\text{minimize } w^T P^T P w,$$
$$\text{subject to } w \geq 0, \quad e^T w = 1. \tag{4.1}$$

Introducing the Lagrange multiplier λ, we state the equivalent special linear complementarity problem of finding $w, v \geq 0$ so that

$$v^T w = 0,$$
$$P^T P w + e \lambda = v, \tag{4.2}$$
$$e^T w = 1.$$

These conditions are just the necessary conditions of Kuhn and Tucker [11] applied to (4.1), and are, of course, also sufficient for this problem. They have a simple geometric interpretation: Noting that

$$\lambda = w^T e \lambda = w^T v - w^T P^T P w = - X^T X,$$
$$v = P^T X - e X^T X = (P - X e^T)^T X \tag{4.3}$$

(where $X = P w$), we see that the complementarity condition $w^T v = 0$ requires that each point P_j whose weight w_j is positive satisfy $P_j^T X - X^T X = v_j = 0$, so that P_j lies on the hyperplane $H(X)$; and the condition $v \geq 0$ requires all points of P to lie on $H(X)$ or on its far side.

Now let Q be any affinely independent subset of P, and R be the remaining points of P, so that $P = [Q, R]$; and let the vectors e, w, v be correspondingly partitioned so that (4.2) may be written—in tableau form—as

	λ	w_1	w_2	
	0	e_1^T	e_2^T	$= 1$
	e_1	$Q^T Q$	$Q^T R$	$= v_1$
	e_2	$R^T Q$	$R^T R$	$= v_2$

$$\tag{4.4}$$

Let M denote the nonsingular submatrix

$$\begin{bmatrix} 0 & e_1^T \\ e_1 & Q^T Q \end{bmatrix}.$$

The system (4.4) has the simple complementary solution found by setting $w_2 = 0$, $v_1 = 0$, and solving for w_1 and v_2; it is exhibited explicitly by performing a block pivot on M, yielding the tableau (4.5) below where $S = R^T R - [e_2, R^T Q] M^{-1} [e_2, R^T Q]^T$.

$$-1, -v_1 \qquad\qquad w_2$$

M^{-1}	$M^{-1}[e_2, R^T Q]^T$	$= \begin{bmatrix} -\lambda \\ -w_1 \end{bmatrix}$
$-[e_2, R^T Q] M^{-1}$	S	$= \quad v_2$

$$(4.5)$$

The system (4.5) is almost, but not quite, that of the "fundamental problem" of complementary pivot theory [6, 12], but it is easy to see how to bring it to that by discarding the first row of (4.5), separating out the first column, and changing signs appropriately to obtain a system of the form

$$q + N \begin{bmatrix} v_1 \\ w_2 \end{bmatrix} = \begin{bmatrix} w_1 \\ v_2 \end{bmatrix}.$$

It will be more convenient for us to stay with the equivalent form (4.5).

Let us number the rows and columns of the tableau (4.5) $0, 1, 2, \ldots, m$, and denote entry i, j of the tableau by $T(i, j)$. The (variable) set of indices $I_Q = \{i_1, \ldots, i_k\} \subseteq \{1, 2, \ldots, m\}$ will designate an affinely independent subset $Q = \{P_{i_1}, \ldots, P_{i_k}\}$ of P. With I_Q is associated the tableau of the form (4.5) (except for irrelevant simultaneous permutation of rows and columns) obtained from the starting tableau (4.4) by pivoting on the block whose row and column indices are $\{0, i_1, \ldots, i_k\}$. The first row and column are special; once having been pivoted in, they are not used again for pivot choices.

Each entry of the tableau (4.5) has a geometric meaning for our problem. We are particularly interested in the left-hand column which gives the values of the basic variables that determine the choice of pivots, and in the diagonal on which we do all our pivoting. What we need is covered in the five propositions below.

I_R will denote the complement of the index set I_Q in the set $\{1, 2, \ldots, m\}$.

Proposition 1. $T(0, 0) = -|X|^2$, where $X = Q w_1$ is the point of smallest norm on $A(Q)$.

Proof. The relations (3.2), with $Y = 0$, follow from (4.4) with $w_2 = 0$ and $v_1 = 0$; and see (4.3).

Proposition 2. For $i \in I_Q$, $-T(0, i)$ is the barycentric coordinate of X in Q.

Proof. Immediate—see (4.5).

Proposition 3. For $i \in I_R$, $-T(0, i)$ is $|X|$ times the distance of P_i from $H(X)$, taken positive if P_i is on the far side of $H(X)$.

Proof. See (4.3, 4.5).

Proposition 4. For $i \in I_R$, $T(i, i)$ is the square of the distance of P_i from $A(Q)$.

Proof. Let $r = P_i \in R$. The column $M^{-1}[1, r^T Q]^T$ from the upper right-hand block of (4.5) is the solution (μ, u^T) of the equations (2.2) for $Y = r$, so that $Q u$ is the nearest point to r on $A(Q)$. Now (2.2) gives $\mu = u^T e \mu = u^T(Q^T r - Q^T Q u) = r^T Q u - |Q u|^2$, so that

$$\begin{aligned}
T(i, i) &= r^T r - (1, r^T Q) M^{-1} (1, r^T Q)^T \\
&= r^T r - (1, r^T Q)(\mu, u^T)^T \\
&= r^T r - \mu - r^T Q u = r^T r - 2 r^T Q u + |Q u|^2 \\
&= |r - Q u|^2.
\end{aligned}$$

Proposition 5. For $i \in I_Q$, $T(i, i)$ is the reciprocal of the square of the distance of P_i from $A(Q \setminus \{P_i\})$.

Proof. Consider the tableau $\bar{T}$ of the form (4.5) formed for the affinely independent set $Q\{P_i\}$. By Proposition 4, $T(i, i)$ is the square of the distance of P_i from $A(Q \setminus \{P_i\})$, and that is not zero. The tableau $\bar{T}$ is obtained from T by pivoting on i, i, so that $T(i, i) = 1/\bar{T}(i, i)$.

5. The algorithm in tableau form

We describe the algebra of the algorithm in tableau form, paralleling the geometric description in Section 2. The notes that follow the algorithm explain the significance of the algebraic steps.

Recall that pivoting on the entry I, J in a tableau T yields the tableau $\bar{T}$ by these rules:

$$
\begin{aligned}
\bar{T}(i, j) &= T(i, j) - t(I, j)\, T(i, J)/T(I, J) \quad \text{for } i \neq I, j \neq J; \\
\bar{T}(I, j) &= T(I, j)/T(I, J) \quad \text{for } j \neq J; \\
\bar{T}(i, J) &= -\,T(i, J)/T(I, J) \quad \text{for } i \neq I; \\
\bar{T}(I, J) &= 1/T(I, J).
\end{aligned}
$$

When the tableau is bordered by variables as in (4.4, 4.5), the variables associated with row I and column J are interchanged, and their signs are changed. We do only *principal* pivots here—always $I = J$—so we always have a complementary basis.

Algorithm
Step 0. (a) Form the $(m + 1)$-by-$(m + 1)$ tableau

$$
T_0 = \boxed{\begin{array}{cc} 0 & e^{\mathrm{T}} \\ e & P^{\mathrm{T}}P \end{array}}.
$$

(b) Let $i = i_0$ minimize $T_0(i, i)$ for $i > 0$. Set $I_Q = \{i_0\}$ and let w be the vector of length m for which $w_i = \delta_{ii_0}$.
(c) Pivot in T_0 on (i_0, i_0) and then on $(0, 0)$; call the resulting tableau T.
Step 1. (Begins with an index set I_Q, an m-vector w, and a tableau T; I_R is the complement of I_Q in $\{1, \ldots, m\}$.)
(a) If $T(0, i) \leq 0$ for all $i \in I_R$, stop.
(b) Otherwise choose $i \in I_R$ so that $T(0, i)$ is maximal.
Replace I_Q by $I_Q \cup \{i\}$.
Step 2. (Requires the above data, and i from (1b) or (3a).)
(a) Replace T by the result of pivoting in T on $\{i, i\}$.
(b) Let y be the m-vector: $y_j = T(0, j)$ for $j \in I_Q$, $y_j = 0$ for $j \in I_R$.
(c) If $y_j > 0$ for all $j \in I_Q$, set $w = y$ and return to step 1. Otherwise, do step 3.

Step 3. (a) Let

$$\bar{\theta} = \min \left\{ \frac{w_j}{w_j - y_j} : \quad w_j - y_j > 0 \right\},$$

and let i be an index for which this minimum is assumed.

(b) Replace w by $(1 - \bar{\theta}) w + \bar{\theta} y$.

(c) Replace I_Q by $I_Q \backslash \{i\}$.

(d) Go to step 2.

Explanation. Step 0 above is clearly the same as step 0 of Section 2. Step (1a), by Proposition 3 of Section 4, stops the algorithm when no point of P is on the near side of $H(X)$; otherwise, step (1b) chooses such a point, and adjoins it to the corral Q.

The index i for step (2a) has been determined either by (1b) or (3a). In the former case, since $T(0, i) \neq 0$, we know that $P_i \notin H(X)$ (Proposition 3), so $P_i \notin A(Q)$, so $T(i, i) \neq 0$ (Proposition 4); and in the latter case $T(i, i) \neq 0$ also (Proposition 5). Thus the pivot operation may be carried out. Step (2b) determines the barycentric coordinates y of the point $Y = Py$ of smallest norm in the affine hull of the new corral Q. If Y is interior to $C(Q)$, then (2c) sets X to Y and returns to step 1, completing the major iteration; otherwise it goes to step 3 for minor iterations.

Step (3a) finds the largest value $\bar{\theta}$ of θ for which the barycentric coordinates $(1 - \theta) w + \theta y$ of the point $(1 - \theta) X + \theta Y$ stay non-negative, and the index i of a coordinate which vanishes for that value. Steps (3b, 3c) replace X by $Z = (1 - \bar{\theta}) + \bar{\theta} Y$, which lies in the face of $C(Q)$ opposite P_i, and designate the new Q as $Q \{P_i\}$.

The argument that $|X|^2 = -T(0, 0)$ decreases monotonically throughout the procedure, and that the procedure is finite, has been given in Section 2.

6. Solution of the example

We shall solve the example of Section 2, giving the result of taking each of the steps of the algorithm. The data are

$$P = \begin{bmatrix} 0 & 3 & -2 \\ 2 & 0 & 1 \end{bmatrix}.$$

Steps		Result

0a

$$\begin{array}{cccc} 0 & 1 & 1 & 1 \\ 1 & 4 & 0 & 2 \\ 1 & 0 & 9 & -6 \\ 1 & 2 & -6 & 5 \end{array}$$

0b $\quad I_Q = \{1\}, \qquad w = (1, 0, 0)$

0c

$$\begin{array}{cccc} -4 & 1 & -4 & -2 \\ 1 & 0 & 1 & 1 \\ 4 & -1 & 13 & -4 \\ 2 & -1 & -4 & 5 \end{array}$$

1a $\quad T(0, I_R) = \{4, 2\}; \qquad$ go on
1b $\quad i = 2, \qquad I_Q = \{1, 2\}$

2a $\quad \frac{1}{13} \times$

$$\begin{array}{cccc} -36 & 9 & 4 & -42 \\ 9 & 1 & -1 & 17 \\ 4 & -1 & 1 & -4 \\ 42 & -17 & 4 & 49 \end{array}$$

2b $\quad y = (9/13, 4/13, 0)$
2c $\quad w = (9/13, 4/13, 0)$
1a $\quad T(O, I_R) = \{42/13\}; \qquad$ go on
1b $\quad i = 3, \qquad I_Q = \{1, 2, 3\}$

2a $\quad \frac{1}{49} \times$

$$\begin{array}{cccc} 0 & -21 & 28 & 42 \\ -21 & 26 & -9 & -17 \\ 28 & -9 & 5 & 4 \\ 42 & -17 & 4 & 13 \end{array}$$

2b $\quad y = (-3/7, 4/7, 6/7)$
3a $\quad \bar{\theta} = 21/34, \qquad i = 1$
3b $\quad w = (0, 8/17, 9/17)$
3c $\quad I_Q = \{2, 3\}$

Steps, contd.   Result

$$
\text{2a} \quad \tfrac{1}{26} \times
\begin{array}{|rrrr|}
\hline
-9 & 21 & 11 & 15 \\
-21 & 49 & -9 & -17 \\
11 & 9 & 1 & -1 \\
15 & 17 & -1 & 1 \\
\hline
\end{array}
$$

2b $y = (0, 11/26, 15/26)$
2c $w = (0, 11/26, 15/26)$
1a $T(0, I_R) = \{-21/26\};$ Stop

Check of solution: $e^T w = 1.$
$$X = P w = (3/26, 15/26);$$
$$\min\{P^T X\} = \min\{30/26, 9/26, 9/26\} = 9/26 = |X|^2.$$

7. Computational notes

A computer routine for a programming problem must guard against having the decisions it makes in the course of its work led astray by roundoff error. The crucial decision in the algorithm of Section 5 is at step (1a): the test there might have stopped the algorithm if perfect arithmetic had been done, but in practice, allow (1b) to choose a point P_i which is actually dependent on Q; and disaster would ensue. We mention here the refinements we have incorporated which seem to make the procedure work well using 16-digit floating-point arithmetic.

Step (1a). Let
$$\phi = \max\{T(0, i)/|P_i| : i \in I_R\}.$$

Stop if $\phi \leq 10^{-10} \sum_{j \in I_Q} w_j |P_j|.$

Step (1b): Choose i as the index for which max above is assumed.
Step (2b): Normalize the y found; replace it by $y/\sum_i y_i.$
Step (3b): Replace all components of w less than 10^{-10} by zero, and normalize w.

The rule for (1a) is based on a simple-minded analysis of roundoff error in the calculation. It has, at least, the virtue, used with (1b), of rendering the choices made independent of the scaling of P. One might also check the size of the pivot entry in step (2a) as a further guard against the dependence of P_i on Q (or on $Q\{P_i\}$ in a minor iteration), but we have not found that necessary.

We have tested the algorithm on two types of generated problem. The two tables below give the average number of pivot steps taken to solve ten problems for various m and n. A type 1 problem is easy: a point X^0 is chosen at random from a cube of side 4 centered at the origin, and the m points of P taken at random from the cube of side 2 centered at X^0 (all distributions uniform). The relative ease of the problem is signalled by the fact that for $n = 10$ the number of points in the terminal corral was usually 3 or 4. A type 2 problem consists of m points, uniformly distributed in the parallelopiped

$$0.995 \leq x_1 \leq 1.005, \qquad -1 \leq x_j \leq 1 \quad \text{for } j = 2, \ldots, n.$$

Such a problem is quite hard, as Table 1 shows. The number of points in the terminal corral was always very close to the smaller of m and n.

Table 1
Average number of pivot steps

n	m	Type 1 20	40	Type 2 20	40
10		1.4	3.1	14.7	28.0
20		3.3	4.3	14.2	34.5
30		4.3	4.1	14.3	30.0

We must mention that the algorithm of this paper, with the calculations performed in a tableau, is probably efficient in arithmetic only for problems in which the number m of points is not much greater than the dimension n of the space in which they reside. Almost all of the calculating is done in the pivot step, which requires $O(m^2)$ multiplications. When $m \gg n$, a different organization of the calculation [22], related to the present method as the revised simplex method for linear programming is to the standard form, is much better; it requires $O(m \, n)$ multiplications for a step, can take advantage of sparsity of the matrix P, and gives better control over roundoff error.

The preceding paragraph explains why we have not much investigated the use of the tableau algorithm in practice. Doing so would involve at least comparing its behavior on a variety of problems with that of other tableau procedures for quadratic programming, such as those of Cottle and Dantzig [5] and Lemke [12], which we have not done. We have, though, tried their procedures on the problem of Section 2 and another.

Counting steps from the tableau 0c of Section 6, our procedure ("T") and the Cottle-Dantzig ("CD") took the same first step, then different second and third steps to solve the problem. The Lemke procedure ("L") requires a special first step (pivoting on 2, 5 after augmenting the tableau by the column e), and then solves the problem in three steps (by reducing to zero the variable associated with the added column).

Next we carefully adjoined the point $(-2, 3)$ to the three points of the problem and reran the methods. (The tableau 0c becomes bordered by 2, 1, -8, 3 on the right and -2, -1, -8, 3, 5 below.) The choices made by T and L were unchanged, so they required, respectively, three and four steps. CD, however, took five steps, because it had to maintain the nonnegativity of the new variable v_4, while L was unaffected and T ignored its change of sign. This suggests to us that T would usually require fewer steps than CD: We have observed in linear programming problems that one should make no effort to prevent reduced costs from changing sign, but concentrate on reducing the objective; and procedure T adheres to that view.

References

[1] E.W. Barankin and R. Dorfman, "Toward quadratic programming", Rept. to the Logistics Branch, Office of Naval Research (January, 1955), unpublished.

[2] E.W. Barankin and R. Dorfman, "On quadratic programming", *University of California Publications in Statistics* 2 (1958) 285–318.

[3] E.M.L. Beale, "Numerical methods", Parts ii-iv, in: *Nonlinear programming*, Ed. J. Abadie (North-Holland, Amsterdam, 1967) pp. 143–172.

[4] M.D. Canon and C.D. Cullum, "The determination of optimum separating hyperplans I. A finite step procedure", RC 2023, IBM Watson Research Center, Yorktown Heights, N.Y. (February, 1968).

[5] R.W. Cottle, "The principal pivoting method of quadratic programming", in: *Mathematics of the decision sciences*, Part I, Vol. 11 of Lectures in Applied Mathematics, Eds. G.B. Dantzig and A.F. Veinott (A.M.S., Providence, R.I., 1968) pp. 144–162.

[6] R.W. Cottle and G.B. Dantzig, "Complementary pivot theory", in: *Mathematics of the decision sciences*, Part I, Vol. 11 of Lectures in Applied Mathematics, Eds. G.B. Dantzig and A.F. Veinott (A.M.S., Providence, R.I., 1968) pp. 115–136.

[7] G.B. Dantzig, *Linear programming and extensions* (Princeton University Press, Princeton, N.J., 1963).

[8] G.B. Dantzig and A.F. Veinott, Eds., *Mathematics of the decision sciences*, Part I, Vol. 11 of Lectures in Applied Mathematics (A.M.S., Providence, R.I., 1968).

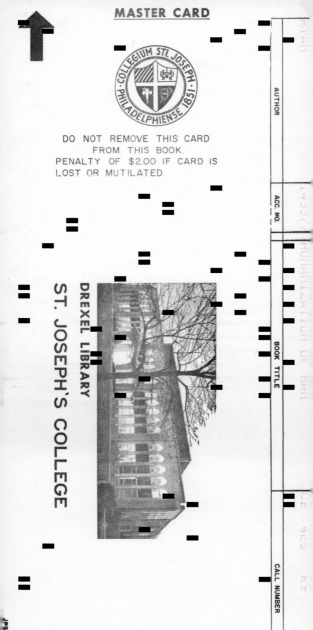

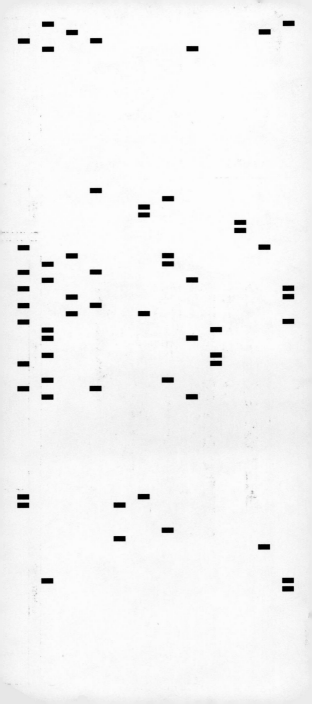